D1379680

THE CLINICAL ENCOUNTER:
THE MORAL FABRIC OF THE PATIENT-PHYSICIAN RELATIONSHIP

PHILOSOPHY AND MEDICINE

Editors:

H. TRISTRAM ENGELHARDT, JR.

Center for Ethics, Medicine, and Public Issues, Baylor College of Medicine, Houston, Texas, U.S.A.

STUART F. SPICKER

University of Connecticut Health Center, Farmington, Connecticut, and National Science Foundation, Washington, D.C., U.S.A.

VOLUME 14

THE CLINICAL ENCOUNTER

The Moral Fabric of the Patient-Physician Relationship

Edited by

EARL E. SHELP

Center for Ethics, Medicine, and Public Issues,
Baylor College of Medicine, and
Institute of Religion, Houston, Texas, U.S.A.

D. REIDEL PUBLISHING COMPANY

A MEMBER OF THE KLUWER  ACADEMIC PUBLISHERS GROUP

DORDRECHT / BOSTON / LANCASTER

Library of Congress Cataloging in Publication Data
Main entry under title:

The clinical encounter.

(Philosophy and medicine; v. 14)
Includes bibliographies and index.
1. Physician and patient. 2. Medical ethics. I. Shelp, Earl E., 1947- . II. Series. [DNLM: 1. Ethics, Medical. 2. Physician-Patient relations W3 PH609 v. 14 / W 62 C641]
R727.3.C53 1983 174'.2 83-17690
ISBN 90-277-1593-9

Published by D. Reidel Publishing Company,
P.O. Box 17, 3300 AA Dordrecht, Holland.

Sold and distributed in the U.S.A. and Canada
by Kluwer Academic Publishers,
190 Old Derby Street, Hingham, MA 02043, U.S.A.

In all other countries, sold and distributed
by Kluwer Academic Publishers Group,
P.O. Box 322, 3300 AH Dordrecht, Holland.

Printed in The Netherlands.

TABLE OF CONTENTS

SECTION III / CONCEPTUAL AND THEORETICAL ANALYSES

SECTION IV / MORALITY IN THE PATIENT-PHYSICIAN RELATIONSHIP

EARL E. SHELP

INTRODUCTION

The encounter between patient and physician may be characterized as the focus of medicine. As such, the patient-physician relationship, or more accurately the conduct of patients and physicians, has been the subject of considerable comment, inquiry, and debate throughout the centuries. The issues and concerns discussed, apart from those more specifically related to medical theory and therapy, range from matters of etiquette to profound questions of philosophical and moral interest. This discourse is impressive with respect both to its duration and content. Contemporary scholars and laypeople have made their contribution to these long-standing discussions. In addition, they have actively addressed those distinctively modern issues that have arisen as a result of increased medical knowledge, improved technology, and changing cultural and moral expectation. The *concept* of the patient-physician relationship that supposedly provides a framework for the conduct of patients and physicians seemingly has taken on a life of its own, inviolable, and subject to norms particular to it.

The essays in this volume elucidate the nature of the patient-physician relationship, its character, and moral norms appropriate to it. The purpose of the collection is to enhance our understanding of that context, which many consider to be the focus of the entire medical enterprise. The contributors have not engaged in apologetics, polemics, homiletics, or empiricism. Neither have they addressed the many individual moral problems in modern health care, nor have they explicated fully the respective rights and duties of the parties. Rather, each from the perspective of his or her own discipline explores the relationship, its conceptualizations, its moral qualities, and its moral standards of conduct. A basic question appears to be either explicitly or implicitly present in almost every essay. Seemingly, each contributor in one way or another cannot escape the quandary about the proper ends of medicine and how those ends may be achieved or approximated in relation to individuals and the moral community. This recurring concern suggests that efforts to understand the patient-physician relationship, to conceptualize it, and to explicate its morality will require a prior consideration of the philosophy of medicine in particular and of moral theory in general. Perhaps, then, the *concept* of the patient-

Earl E. Shelp (ed.), The Clinical Encounter, vii–xvi.

physician relationship and its moral force will be better understood. The essays in this volume are efforts toward this end.

The essays are presented under four general headings. The first section contains three historical essays charting the evolution of the patient-physician relationship and the moral norms associated with it. The second section contains two series of exchanges between several authors on 'Models of the Patient-Physician Relationship'. Five conceptual analyses follow in the third section, and three explorations of the morality of the encounter between patients and physicians are provided in the concluding section. These inquiries and exchanges seek to enrich our understanding of the concept and the phenomenon it represents. In addition, they suggest other apparently overlooked areas of investigation that might be explored with profit.

The historical section opens with a sketch by Darrel Amundsen and Gary Ferngren of some significant developments within the patient-physician relationship from Antiquity through the Renaissance. During the examined period there was a variety of healing sources available who represented divergent ideologies and engaged in practices consistent with their respective belief systems. The variety of healing sources present during these centuries and the nature of the relationship between physician or healer and patient, particular to each, reflect the multiple paths that have been followed to effect cure or bring relief to the sick. The evolution of the patient-physician relationship and the distinctions of the different sorts of relationships are reviewed by the authors along several lines of inquiry: notions of disease causation, status of physicians, role of philanthropy, attitudes toward physicians, duties of physicians, and considerations of medical etiquette. They suggest that the modern concerns regarding knowledge, professional etiquette, motivation, and technical competence have precedent in the evolution of medicine and the contact of healers with sick people from primitive cultures to the Renaissance.

Laurence McCullough extends the historical inquiry into the 18th and 19th centuries. He probes the Anglo-American history of medicine during this period to correct what he considers to be basic misperceptions about its ethical legacy, expecially with respect to the morality of the patient-physician relationship. He argues that during the 18th century there is evidence of a genuine medical ethics in addition to the much-discussed medical etiquette as found in the work of Thomas Percival and the first code of the American Medical Association. The work of Scottish moral philosophers, such as Francis Hutcheson, that focused on notions of benevolence and duty provided a foundation for a patient-centered account of the ethics of the patient –

physician relationship represented in the writings of Samuel Bard. This beneficence-based medical ethics of the 18th century was displaced in the 19th century. Intraprofessional concerns and discordant views of the telos of medicine contributed to a severance of medical ethics from its ground in moral philosophy and a substitution of self-interest for philosophical reasoning in medical reflection. McCullough thinks that it is not technology which confounds contemporary medical ethics. Rather, the fundamental problem is the presence of a plethora of moral understandings of technology and the presence of many views regarding the telos of medicine that previously helped to define what was proper to it. He concludes by suggesting three lessons that contemporary medical ethicists can learn from the study of their forebears.

A review of American codes of medical ethics and their effect on the patient-physician relationship is advanced in an essay by John Duffy, which concludes the historical section of the volume. He cautions that such an inquiry requires a separation of ethics from economics and a differentiation of medical etiquette and medical ethics. Given these difficulties, Duffy reviews the provisions of the 1847 code of The American Medical Association and its subsequent revisions until 1957 when it was replaced by a 'Principles of Medical Ethics'. He thinks that the professional struggles for greater public approval and economic security reflected in these statements had little effect on individual physician-patient relations or the image of the medical profession. He suggests that many of the moral issues in medicine in the 20th century are not addressed by the American Medical Association codes. Instead of matters of morality, matters of professional etiquette and the economic welfare of the profession are addressed. Thus it seems that, if Duffy is correct, codes and statements of principles by professional organizations like the American Medical Association are of little help in defining a morality for the patient-physician encounter or in resolving the seemingly countless moral disputes about the practice of medicine.

The second section of the volume consists of two series of exchanges on 'Models of the Patient-Physician Relationship'. K. Danner Clouser initiates the discussion with a critique of the relationship of models to matters related to ethics in previously published essays by Robert Veatch [4] and William F. May [3]. With respect of Veatch's models of the patient-physician relationship (Engineering, Priestly, Collegial, Contractual), Clouser asks, What is the point? – since their relation to ethics is not made clear. He thinks that it would be better to dispense with models and their implications for the moral practice of medicine in favor of a clear statement of what would be immoral for a physician to do to a patient.

Clouser finds a link between Veatch and May: The concern to find a format that would most likely maximize morality and motivate moral conduct. He sees May's dissatisfaction with code and contract models as their failure to express May's notions of proper beliefs, attitudes, and motives on the part of physicians that generate right conduct. Clouser thinks that *models* may inspire motivation and a philosophy of life but, as such, proper motivation is not held to be necessary or sufficient for moral conduct. What is moral is logically independent of and determined by criteria other than one's personal philosophy.

Clouser concludes his critique with a sketch of his view of ethics as rule-based – centered around the theme of 'do not cause evil'. A feature of this proscription is a prescription to do one's duty, since failure to do so results in an evil. He sees professional codes as delineating the nature and substance of professional duty. The content of these codes must be in accord with other basic moral rules. The use of models to clarify the patient-physician relationship is seen by Clouser as obscuring rather than clarifying the relevant moral issues, i.e., deeds and duties. He argues that the patient-physician relationship carries the same basic moral obligations as all other human relationships. Additional duties particular to this relationship are subject to identification in professional codes.

Robert Veatch responds to Clouser's critique with 'The Case for Contract in Medical Ethics'. Veatch agrees that the normative content is the core of medical ethics but argues that the basic work of metaethics must not be neglected. He identifies two variables that were part of a theoretical framework behind his earlier discussions of models: (a) the necessity of professional value judgments, and (b) the convergence of lay and professional values. He defends the use of models in this task as pointing toward fundamental problems to be resolved in any fully developed theory of medical ethics.

Clouser replies to Veatch's explanation but finds it unconvincing. He does not alter his criticism that the relation of models to ethics has not been shown.

The second exchange is between Baruch Brody, Patricia White, and Isaac Franck. Here the subject is legal models and their inherent claims for the patient-physician relationship. Brody discusses a *commercial contract* model that places emphasis on consent to the relationship and a *status* model, that places emphasis on the content of the relationship. He sees the commercial contract model presupposed in common law and the status model presupposed in Judaic law. He explores the meaning of the claims represented by each model for physicians and for patients. The apparent values, strengths and weaknesses of each model are examined. Brody questions the use of either model in establishing the proper relation between patients and physicians since both fail to capture the *sui generis* nature of the patient-physician relation. Of the two, however, he sees the status model as more capable of accommodating the unique nature of the relationship and its social context.

Responding to Brody's remarks concerning the common law, Patricia White agrees that it fails to capture adequately the morality of the patient-physician relationship. But she suggests that it does a better job of doing so than Brody allows. Tort law should be considered in addition to contract law if one is to assess the adequacy of the common law's recognition of the *sui generis* nature of the relationship. Yet, she counsels that legal standards are not necessarily the same as moral standards and neither are they particularly helpful in constructing a fully adequate moral model of the patient-physician relationship.

A second response to Brody is provided by Isaac Franck, who focuses on Brody's understanding of the patient-physician relation as represented in Jewish law and the extent to which Brody's status model is represented in it. Professor Franck disagrees that Jewish law in its 'full authenticity' discusses the patient-physician relation by analogy to some other non-medical relation. The *sui generis* nature of this relationship is represented as a unique covenantal, triadic relation between patient, physician, and God, unlike any other practitioner-client relation, with notions of responsibility guiding the conduct of the parties. Franck finds evidence for this claim in the reasoning behind orders of Jewish courts compelling physicians to care for the poor without charge and in counsels regarding the bases for remuneration. He leaves open the question of the extent to which the status model and the concept of social order inherent in it – reflected in the 'full authenticity' of Jewish law – can be implemented in the democratic West.

Brody briefly responds to Professors Franck and White. He agrees with Franck that from a Jewish perspective God is a party to all legitimate relations. But, Brody disagrees that this and the other points raised by Professor Franck establish that Judaism treats the relation between patient and physician in a *sui generis* fashion. With respect to Professor White's comments, Brody disputes her contention that the common law treats the patient-physician relation any different from other contractual relations.

The third section of essays turns from discussions of models of the patient-physician relationship to analyses of the relationship to discover what distinguishes it from other human associations, what norms are appropriate to it, and what defines its boundaries. Edmund Pellegrino suggests that the central dilemma of modern medicine is to construct a philosophy of medicine that in turn will shape the relationship of physicians to patients and to society. He seeks an organizing principle whereby the boundaries of medicine can be set. He examines notions of medicine as a body of knowledge, as defined in terms of its ends, and as context dependent as a prelude to proposing an organizing principle that is empirical and phenomenological. He argues that the nature of the healing relationship is the distinctive feature of clinical medicine and that its immediate end of healing serves as its architectonic principle. The notion of healing as an architectonic principle is then explored for its implications for medical practice and medical ethics.

The psychiatric patient-physician relationship is the subject of Joseph Margolis's contribution. He focusses primarily upon *Diagnostic and Statistical Manual* classifications [1] and the recent work by Charles Culver and Bernard Gert [2] to clarify some ways in which the, at times, intersecting interests of individuals, intermediate social groups, and society affect psychiatric theory and the psychiatrist-patient relationship. He identifies and explores a series of conceptual uncertainties that revolve around the social responsibility of psychiatry as a medical profession. Margolis discusses the nature of the range of complaints brought to psychiatrists and the ideological, pluralistic, and relativistic dimensions of psychiatric knowledge and practice. In light of this, he defends a view that there are at least three zones in which norms of psychiatric practice are found: medical-psychiatric, social, and client-centered. The norms of each zone have implications, according to Margolis, for understanding the full texture of the psychiatrist-patient relationship.

The changing nature of the relationship between patient and physician in general is examined by Robert Veatch in the next essay. Professor Veatch explores the difference it makes in medical ethical theory whether the patient and physician relate as strangers or as friends. He contrasts 'Hippocratic Theories of Medical Ethics' with 'Contract Theories of Medical Ethics' to determine which more adequately accommodates the way in which health care is provided in industrialized societies. He favors a Contract Theory since, in his view, it can accommodate more easily both the friendship model and stranger model of the therapeutic relationship, even though the framework for each is different.

John Ladd is also concerned with how patients and physicians relate to one another. In "The Internal Morality of Medicine', Ladd moves away from a 'model' approach to a 'problem' approach as a means of defining the aims of medicine and the principles of the internal morality of medicine. The internal morality of medicine is understood as those moral norms in a clinical situation that depend on medical considerations. It is this internal morality to which physicians look for guidance, says Ladd, whereas patients appeal to moral concepts and norms external to medicine. Hence, there is potential for conflict between the parties. Ladd explicates his preferred 'problem' approach with respect to how it defines the aims of medicine, and shows how it can be applied to clinical medicine so as to explain some of the conflict within patient-physician relationships. He suggests that these conflicts may be negotiated through a dialectical process.

The final essay in the conceptual and theoretical section is by George Agich who discusses the scope of the *therapeutic relationship* as a *conceptual* question. He examines the nature of medicine as a practice governed by rules that are either written or unwritten. It is also a practice set against a background of beliefs and values that renders the rules intelligible and compelling. The disputes with regard to the scope of the therapeutic relationship are usually expressed in terms of *rights* and *duties* even though what is at issue, at times, are the operative values and beliefs that are not explicit. The therapeutic relationship can assume many cultural forms of which the patient-physician relationship is only one. Agich explores several models of therapy in medicine as a means to display the structure of therapeutic relationships and the place definitions of 'well-being' and 'power' have in them. He concludes that conceptual analyses taken alone – of medicine as a practice, of a therapeutic relationship, of models of therapy, or the parameters of therapy – cannot adequately define the scope of therapy in medicine. He urges that these pursuits are points of departure but that the task will require also a normative justification of the beliefs, values and rules relevant to therapy in the practice of medicine. Beyond this, a consideration of the content and structure of medical practice in particular settings would be necessary.

The fourth and final section of essays explores the moral features of the patient-physician relationship. H. Tristram Engelhardt, Jr., echoes concerns voiced by other contributors in his essay on 'The Patient-Physician Relationship in a Secular, Pluralist Society'. Engelhardt views society as secular to the extent that policies and rules are framed in language that does not appeal to religious authority. It is pluralist to the extent that divergent moral viewpoints are represented. Thus, the patient and physician may come together as moral strangers, each with allegience to different moral communities, but sharing a concern about a specific medical decision. He sees health care occurring within such an environment by developing notions of a peaceable community. Bioethical controversies in such a setting may not be resolved, but patients and physicians nevertheless can relate on the basis of mutual respect. He defends the procedural safeguards of bureaucratic regulations in such a society. These protect individual viewpoints without establishing one as superior by force. The physician functions as a type of bureaucrat and geographer of value assisting patients in making decisions without imposing a particular choice. He suggests that an approach to these matters like the one he proposes enables an individual to be a part of particular moral communities and to be a part of a broader community based on mutual respect.

Albert Jonsen speaks to physicians and patients rather than philosophers as he considers the necessity of moral conduct within the therapeutic relationship. He attempts to unravel the importance of good morality to good medicine. Jonsen suggests that moral conduct is intrinsic, and thus necessary, to 'good' medicine. He describes the therapeutic relationship in terms of its 'moral archeology'; altruism and self-interest, expectations of the parties, diagnostic and therapeutic practices, effect on the social environment, and the moral probity of private behavior. He proposes that respect, honesty, courage, and justice are virtues appropriate to the therapeutic relationship. Jonsen concludes that the relevance of these virtues provides evidence of the necessity of moral conduct for the therapeutic relationship.

The collection ends with an essay by David H. Smith that provides a sketch of the conceptual world of Christianity in order to suggest ways in which such a perspective can influence the structure and process of the therapeutic relationship. He discusses Christian themes of monotheism, the kingdom of God and selected norms of human citizenship within it (especially those derived from the postulate of the sovereignty of God over life and death), explanations of human suffering in relation to Christological affirmations, and bonding. Each theme is shown by Smith to have relevance to the patient-physician relationship. These themes are held to provide an interpretive framework that is both moral and epistemological. As such, the author believes that they have implications for the relationship between patient and physician.

It is hoped that these essays will not only inform the contemporary discussions of the patient-physician relationship but will stimulate a more fundamental investigation of the concept and phenomenon than has been undertaken previously. The paucity of scholarly literature on the concept, its moral force, and the theoretical significance of the patient-physician relationship suggests that these concerns have been explored only superficially. Yet it seems that much could be gained by additional studies that explicate the concept and place it within a more fully developed philosophy of medicine.

Edmund Pellegrino is probably correct in his assessment that this is the single most significant challenge facing the participants in these sorts of inquiries. Perhaps order and consensus can be brought to what seems to be chaos and endless dispute in bioethics by undertaking the development of coherent theories that will provide a structure within which individual concerns can be placed. The essays in this volume suggest that such a project appears worthwhile. The contributors seem unable to consider the patient-physician relationship without considering, as well, concerns broader in scope. Yet they seem to feel that both in the past and at present discussions about the patient-physician relationship and about the conduct of the parties to it cannot, and perhaps should not, escape evaluation by standards other than those internal to it. They do not deny the importance of these internal standards but they seem not to be certain that these standards alone are sufficient. Thus a discovery of more adequate ways to conceptualize medical practice, especially at the point of interaction with patients, is indicated. The essays in this volume have identified several ways in which this task may be approached. They provide a point of departure for others interested in these matters to consider, correct, adopt, reject, or move beyond. Moral discussions of decisions and conduct within the patient-physician relationship are important. But so is the task of conceptualization and theory formation. It is hoped that these essays will promote this sort of critical and systematic inquiry.

A volume such as this is the product of a collaborative effort of many individuals. The good will, generosity, and commitment of each is required in order for it to be completed. Except perhaps for those rewards that are intellectual in nature, the compensation for being involved in a project like this is minimal. Because of their charitableness and because of the integrity of their work, as editor I express my appreciation to them all, individually and collectively. But to stop here in displaying one's debt and gratitude would be negligent since several others whose contribution does not appear in the Table of Contents would go unrecognized. Mrs. Audrey Laymance again has demonstrated her skills as an editorial assistant in the preparation of the manuscript. Without her able assistance this volume would still be incomplete. The editors of the series, H. Tristram Engelhardt, Jr., and Stuart F. Spicker again have provided me with their counsel and good offices in the design and production of the volume. Susan M. Engelhardt carefully reviewed the manuscripts and proofs for errors in production. I also wish to thank the staff at D. Reidel Publishing Company who should know of my appreciation for their competent assistance at every phase.

Now a personal word of gratitude to Ronald H. Sunderland. Professor Sunderland was the Director of the Institute of Religion until the close of 1982 when he resumed his teaching post full time. I have been privileged to teach on his faculty at the Institute of Religion. I have been honored by his friendship. His leadership and encouragement in part are responsible for this volume. He has been a student of the patient-physician relationship for years in his role in hospitals as an educator in pastoral care. His perceptions of patients and physicians, and the pastoral sensitivities that he brings to their encounter as educator and minister lend dignity and grace to situations that otherwise are unredeeming. So Ron, my friend and colleague, this volume is for you. May you learn from it as I and countless others have learned from you.

October, 1982

BIBLIOGRAPHY

1. American Psychiatric Association: 1980, *Diagnostic and Statistical Manual of Mental Disorders*, 3rd. ed. [DSM–III], Washington, D.C.
2. Culver, C. M. and Gert, B.: 1982, *Philosophy in Medicine: Conceptual and Ethical Issues in Medicine and Psychiatry*, Oxford University Press, New York.
3. May, W. F.: 1975, 'Code, Covenant, Contract, or Philanthropy', *Hastings Center Report* 5 (December), 29–38.
4. Veatch, R. M.: 1972, 'Models for Ethical Medicine in a Revolutionary Age', *Hastings Center Report* 2 (June), 5–7.

SECTION I

HISTORICAL INQUIRIES AND PERSPECTIVES

DARREL W. AMUNDSEN and GARY B. FERNGREN

EVOLUTION OF THE PATIENT-PHYSICIAN RELATIONSHIP: ANTIQUITY THROUGH THE RENAISSANCE

BASIC DEFINITIONS

Etymologically, 'patients' are simply those who suffer or endure something. In common usage, they are those who suffer or experience some perceived ill or dysfunction or injury that causes them to seek help, succor, or relief. It is only when help, succor, or relief is sought and the sufferers enter into a relationship with those whose aid they have requested that they can properly be called 'patients', for the term suggests those who are in a state of need vis-à-vis others who are in a position to render assistance addressed to that need. The word 'patient' is not properly applied to one[1] who has a perceived need in the realm of ill or dysfunction or injury *until* one enters into a relationship with another to whom one turns for aid. Until then one can only be described as a potential 'patient' by virtue of one's need. Nevertheless one can properly be termed a 'patient', even though one suffers or endures nothing for which aid is sought, if one places oneself in a relationship with another who seeks to determine whether an unperceived need exists or who gives advice or assistance that is prophylactic of some potential ill or dysfunction. By convention, the need, whether real, imaginary, or potential, must be physical or (at least in more recent times) psychological, although the term 'patient' may be applied metaphorically to one experiencing a need other than physical or psychological. In the latter sense the one to whom one turns for aid may also metaphorically be called one's 'doctor', 'physician', 'surgeon', or, perhaps, 'therapist'. These terms are etymologically further from their current meaning than is the word 'patient'. Etymologically, a 'doctor' is one who teaches; a 'physician', one concerned with questions of nature; a 'surgeon', one who deals manually; a 'therapist', one who heals, cures, or cares. However, in common usage, a 'doctor' may be a 'physician' or a 'surgeon' or both. 'Physician' may be used interchangeably with 'doctor' but 'surgeon' may not be. 'Therapist', in common parlance, is not a substitute for any of the other three, but denotes one dealing, on a less sophisticated level than they, with the physical (and sometimes psychological) needs of patients. We shall use the terms 'physician' or 'doctor' interchangeably to include 'surgeon' but to exclude 'therapist', and shall consider as a 'physician'

Earl E. Shelp (ed.), The Clinical Encounter, 1–46.

or 'doctor' anyone who is functionally and ideologically recognized as one to whom the potential patient may go, or one whom the patient may call for help, succor, or relief of physical or, perhaps, psychological ill, dysfunction, or injury.

Major conceptual differences exist between the functional and ideological aspects of being a patient and being a physician. Physicians are physicians irrespective of their relationship with patients, and they are so by virtue of their supposed knowledge or skill that renders them capable of extending help, succor, or relief to those in need of their assistance. One would not refer to a physician who does not have, at the moment, any patients as a potential physician, although it is quite proper to refer to anyone, even to the physician, as a potential patient. Thus one is a physician in an absolute sense, while one is a patient only in a relative sense, both temporally and relationally. One is also a patient because of what one needs, while one is a physician because of what one can give as a physician. Thus, in respect to their relationship, the physician is a resource to which the patient turns and is, at least potentially, an authority, while the patient, within the relationship and insofar as the patient is willing, accepts the other's authority as being, at least ostensibly, for the patient's good. Within the relationship the physician is in a position of strength, while the patient is in one of weakness and vulnerability, whether potential or realized. This is true of the patient-physician relationship insofar as it deals with the presence to need on the part of the former and the capacity on the part of the latter of render assistance, because of special knowledge or skill, to that need, even when the patient is the social, economic, legal, or political superior of the physician.

Additionally, patients, although they are those in need, and thus those who seek to receive, are also those in the relationship who give, in that they must usually compensate physicians for the services rendered. By the same token, physicians, although they give or extend their services to patients, are the receivers of a benefit that provides at least part of the motivation for their entering into the relationship, and for obtaining the requisite knowledge and skill, whether the benefit is economic, as is usually the case, or intangible, e.g., personal satisfaction or spiritual reward.

The definitions advanced above seem reasonable for the Western world and we assume in the present historical investigation that these definitions transcend time and place and can be considered basic to the patient-physician relationship. Of course the extent of physicians' authority, strength, and prestige in the relationship will depend in great part on the degree of their specialized knowledge, their social status, and the roles, other than directly medical, that they may occupy in the particular culture that are allied with, or ancillary to, their curing, healing, caring, or prophylactic functions.

PATIENT, PHYSICIAN AND CAUSALITY

Among pre-literate or primitive people, religion and magic are usually one and the same and medicine is subsumed under them. When people experience any physical or psychological ill or dysfunction for which the cause is not readily apparent, they turn to the only members of their society who can determine which power caused the ill or is manifested in it and how it may be removed. Those whose help they seek are almost without exception the most learned men in the community, who possess an intimate understanding of the supernatural as well as the skill to use their knowledge to the advantage of members of the community. They are variously called 'medicine men', 'shamans' or 'witch doctors'. The noted medical historian Henry Sigerist most perceptively observes that "it is an insult to the medicine man to call him the ancestor of the modern physician. He is that, to be sure, but he is much more, namely the ancestor of most of our professions" ([33], Vol. 1, p. 161). He performs most, if not all, his functions through magico-religious means and most of these functions are directed specifically toward preserving or restoring the welfare of the community. The high degree of success in his treatment of disease (which is only one aspect of his diverse responsibilities) is accomplished through incantations, prayers, disease-transfer, dances, and sacrifices, and depends upon his strong belief in the efficacy of the magical process and the patient's and the community's faith in both him and his means. Such an individual is much more than a physician and the dignity, indeed the aura, attached to his person and his office is incomparably greater. One's dependence upon him, while in a state of specific need (illness, disease, injury), is but an extension of one's dependence upon him in the much broader spectrum of stability, prosperity, well-being and even survival, both individual and communal. The 'patient-physician relationship' in pre-literate or primitive societies is thus based upon a dependence-authority set, of which the typical patient-physician relationship in more advanced societies is but a dim reflection. The difference is due, in great part, to the modern conceptions of etiology or causality of disease. In primitive or pre-literate societies, where most diseases are seen as having an exclusively supernatural origin – e.g., demons or spirits acting without discernible motive or in response to the breaking of taboo – only those who understand this realm can deal with the disease and they, by virtue of their knowledge and skill and, very importantly, their willingness to employ them for the individual and the communal good, are worthy of respect and fear and veneration.

Primitive societies, however, do not attribute all disease to magic or possession. When individuals are afflicted with common ailments (e.g., upset stomach, bee-sting, coughs) they or a member of their family or village may treat them empirically without reference to the supernatural. More complicated problems (e.g., dermatological or gastric conditions, as well as wounds and fractures), for which a cause that is not seen as supernatural may be identified, may be treated by a 'herbalist', who is sometimes also the 'medicine man'. In some societies 'herbalists' are not 'medicine men' but are significantly lower than the latter in power (over both the supernatural and the community) and prestige. Whether we should properly speak of 'herbalists' as 'physicians' is debatable since the range of complaints with which they deal would exclude the majority of conditions with which 'physicians' have traditionally dealt and which are within the purview of medicine men in pre-literate societies. On the other hand, it is perhaps more proper to call 'herbalists' rather than 'medicine men' 'physicians' since the 'herbalists' deal with ailments generally without reference to the supernatural, except insofar as the substances they employ are thought to possess a potency that is itself supernatural. We must be careful not to make too fine a distinction here. A particular root or herb may be well-known as having a mysterious power that makes it efficacious for various uses. No great dignity attaches to the man, woman, or child who seeks it out and applies it. If, however, individuals have knowledge of a wide variety of effective substances with which the majority of people are not familiar, they may be respected for the extent of their knowledge. If any of these substances, however, is rendered effective only by incantations or similar means, it is no longer something that can be administered or used simply by those who are knowledgeable of the substance's efficacy, but the services of 'medicine men' are then necessary. This is why 'medicine men' and 'herbalists' are, in some societies, one and the same. When the herbalists are not also medicine men but simply treat ailments with substances that, regardless of how powerful and mysterious they may be, are applied without accompanying spells that only medicine men are able to perform, they remain at a level much closer to that of those whom they are treating. Although they may receive respect and be obeyed, it is by virtue of their knowledge or skill that, although superior to that of their patients, differs in degree but not in essential quality. While the medicine men are, among other things, sorcerers, diviners, priests, healers, and magicians, herbalists are essentially craftsmen, however arcane some of their knowledge may appear to be. The relationship of the former to their 'patients' is on a significantly different level than that of the latter to those whom they treat.[2]

This division, between the supernatural and what may strike us as 'natural' and 'rational' in medical treatment, remains (although not always pronounced) as we move from primitive or pre-literate societies into the earliest civilizations of the ancient Near East, i.e., Egypt and Mesopotamia (Sumerian, Akkadian, Babylonian, Assyrian, and Chaldaean). Persia, as we shall see, is quite distinct in this regard.

Both in Egypt[3] and Mesopotamia,[4] illness and injury were viewed as creating disharmony between the one afflicted and one's total environment. As among pre-literate peoples, disease generally was not viewed symptomatically but etiologically. The cause *was* the disease since the disease itself was the supernatural being or force that had penetrated the afflicted person. Sometimes the gods were seen as the authors of disease and the affliction was regarded as punitive. The linking of sin and illness was not uncommon; e.g., in the Code of Hammurabi a particular deity is said to inflict sickness on those who do not obey the law.

Mesopotamia, more than Egypt, was similar to primitive or pre-literate cultures in having a fairly clear delineation, both functionally and ideologically, in 'medical' services. There were two types of 'physicians', the *āšipu* and the *asû*, who were invariably male. The *āšipu* was an exorcist, magician, and priest. Concerned with diagnosis and prognosis, he sought to identify the demon that caused – or was – the disease, and to determine its intention and whether the illness would be fatal. If his prognosis was unfavorable, he would withdraw from the case. The *āšipu* seldom, if ever, used drugs, but relied on incantations, prayers, libations, and so forth.

The *asû*, by contrast, was not a priest. He was, on the one hand, closer to the primitive herbalist and, on the other, to what we would call a physician and surgeon. He was not concerned with etiology. He was both a pharmacist and prescriber of drugs and employed a wide range of empirical means. He was basically a craftsman who was concerned with therapy addressed to the relief of pressing and acute symptoms. His craft was quite independent of that of the *āšipu* and probably had its roots in earlier primitive empirical medical practices.

When ancient Mesopotamians became ill, they could choose to be treated by either an *āšipu* or an *asû*. In making their decision they would be guided in great part by whether the ailment was one for which, in our terms, a 'natural' cause was obvious. In such a case they would probably directly consult an *asû*. Some probably went first to an *āšipu* and then to an *asû*, either at the *āšipu's* recommendation or because he refused to treat the case. Sometimes an *āšipu* and an *asû* worked on the same patient together.

In Egypt the divisions were not as clear as in Mesopotamia. Those suffering from diseases could avail themselves of the services of priests of a wide variety of deities who were primarily or peripherally concerned with healing. These, however, were not the main sources of medical care. There were three types of medical practitioners, who were also male:

(1) The *sa.u.*, who was a sorcerer and exorcist, primarily employing charms and incantations as well as some drugs.

(2) The *wabw*, who was a priest of Sekhmet, a goddess of healing. These priests apparently combined 'rational' medical or surgical procedures with magico-religious practices.

(3) The *swnw*, who was a physician in roughly the same sense as the *asû* in Mesopotamia and was not a priest but essentially a medical craftsman.

It should be noted that all three of these offices or roles might be combined in one person. In any event, there was no rivalry between the three, just as there was none between the *āšipu* and the *asû* in Mesopotamia. When an Egyptian became ill, the choice of which healing source to employ probably depended in great part on the nature of the disease, i.e., on the perceived etiology. These three types of 'physicians' represent different but complementary systems entirely compatible with each other in cultures of which a magico-religious consciousness was an integral part and in which magical, religious, and 'rational' medicine overlapped. In such cultures those who practiced magico-religious medicine did not enjoy nearly the prestige or possess the power of the medicine man in primitive societies. Thus their authority over their patients and the awe in which they were held were proportionally less. Yet their function was essentially sacerdotal, a fact that significantly tempered their relationship with their patients, especially since they were not priests who employed 'rational' medicine merely as an extension of their sacerdotal office but practiced magico-religious medicine as an essential component of their priestly role.

Those who practiced medicine basically as craftsmen would not have enjoyed the prestige of the magico-religious 'physician', just as their predecessors, the primitive herbalists, were significantly below the medicine men in status and veneration. The gap, however, has narrowed significantly. The specialized knowledge and skill of the medical craftsmen in the ancient Near East probably afforded them a measure of quasi-professional status similar to that enjoyed by anyone then occupying a role for which practical skill was requisite.

Persia presents a far different picture.[5] The national religion was Zoroastrianism, a fundamentally dualistic system, in which Ahuramazda was the spirit of light, the creator of good, and Ahriman the spirit of darkness, the creator of evil. Suffering and disease, as evils, were caused by Ahriman, but joy, health and healing came from Ahuramazda. Within such a religion, the restoration of health was part of the struggle of light against darkness. Thus those engaged in the healing arts were auxiliaries of Ahuramazda, fighting against the forces of Ahriman. The ill, and even the injured, were victims of Ahriman. They were polluted and in need of purification. The act of healing was itself a form of purification and thus all healing arts were essentially purificatory and all healers were priests.

There were three classes of healers, all of whom were priests. They are here listed in order of status:

(1) Those who practiced magico-religious medicine, healing by the holy word.

(2) Those who treated the ill with drugs.

(3) Those who practiced surgery (setting of fractures, cautery, and minor operations).

Because of the peculiarities of Zoroastrian dualism and the unique place of disease and healing within it, Persian medicine had a religious mission that was different, at least in degree, from the role of medicine in Egypt and Mesopotamia. The sacerdotal nature of *all* medical practice rendered the role of any of these three types of physicians in their relation with their patients qualitatively more elevated and authoritative than that of medical craftsmen. Nevertheless, the level of prestige, if not of authority, declines as the medical procedures move from the magico-religious to the pharmacological and finally to the strictly manual.

What of the attitude of 'physicians' toward their 'patients' among primitive or pre-literate people and in the ancient Near East?[6] First of all, we must distinguish between the types of afflictions being treated. The more specifically symptomatic and the less etiological their understanding of a case, the closer they were to modern attitudes. Craftsmen treating, e.g., broken bones, wounds, dermatological complaints that can be alleviated by certain unguents, or gastric problems for which an obvious cause is identifiable, would likely view their patients as those who have experienced a misfortune by accident or uncontrollable mishap, or have, through indiscretion, brought a problem upon themselves. The cause-and-effect relationship was seen as being in the realm of what we would call the 'natural'. As we move into the sphere of disease, we enter that mysterious realm of the supernatural. Generally, those suffering from disease in primitive societies and in the ancient Near East, excluding Persia, were viewed as being in a state of disharmony with their environment, an environment of which the supernatural was a most vital and integral part. At least to an extent, those afflicted with disease were considered polluted. It is important to note, however, that they were not seen as unclean, untouchable, possessed creatures to be shunned. The sacerdotal nature of 'medical' treatment of those afflicted with disease militates against a contempt for the ill by those treating them when the causality is viewed as supernatural. But there was not, at least in Egypt and Mesopotamia, any correlative sense of responsibility either (1) to attempt to treat those for whom no hope of recovery was held out, or (2) to attempt to alleviate the pain and suffering of those afflicted with conditions thought to be incurable.

In Persia, the ill were held to be unclean and were segregated. Given the peculiar situation of ancient Persia, this fact did not lessen the responsibility of the physicians to attempt a cure, but indeed heightened it. "In the Zoroastrian struggle between the forces of good and evil, the physician was clearly aligned with the good, and his patients, when polluted by disease, turned to him, being compelled not only by the motivation shared by the ill in any culture but also by religious necessity. And, significantly, Persian physicians do not appear to have refused to treat hopeless cases" ([4], p. 881).

Among the early Greeks,[7] attitudes toward disease were quite similar to those of the Mesopotamians and Egyptians. The Greeks of Homer's time and of the archaic period (ca. 800–500 B.C.) generally attributed disease to supernatural causes, i.e., *daimones, alastores* and *Keres* (malignant powers or evil spirits), or the gods. When the affliction was viewed as coming from a god, attempts were made to determine the cause of the deities' displeasure and to placate them through sacrifice or purification. There were various healing shrines dedicated to gods or heroes to which the afflicted might go to pray and hope for supernatural healing. There was, however, no priestly class and thus no sacerdotal physicians, as such. In this regard the Greeks of the archaic period were closer to pre-literate peoples than to ancient Near-Eastern civilizations, as they had physician-seers, called *iatromanteis*. These were, in some ways, similar to medicine men, witch doctors, or shamans in their practices, since they used charms, various methods of purification, exorcism, and spells. They were itinerant, travelling from city to city, and employed religious means and magic to turn away pollution and disease.

There were also *demiourgoi*, empirical medical craftsmen, similar to primitive herbalists and their ancient Near-Eastern counterparts, treaters of wounds and minor ailments, who relied on skill, observation, and experience.

As was true of craftsmen generally, their knowledge was passed on to apprentices. By the end of the sixth century B.C., groups of these physicians and their student apprentices began to gather together in various cities in the Greek world.

It was in the fifth century B.C. that the Greek medical craft began to take on the form of a science. While a good deal of empirical technique had previously been accumulated, no effort had been made to formulate a body of theoretical knowledge within which to place empirically-efficacious procedures, to develop theories of health and disease to explain disease in terms of natural causation. During the fifth century B.C., various medical craftsmen turned to philosophy, thinking that they could thereby gain, because of philosophy's attempt at universal formulation, a correct understanding of the nature of man. The Hippocratic Corpus provides our earliest example of this new medicine that developed a theoretical basis for medical practice. Greek medicine became at the same time both rational and empirical, broadening the scope of empirical or craft-medicine to include disease, which had previously been primarily within the purview of those whose competence and knowledge made them capable of dealing with conditions to which a supernatural etiology was attributed. Now medical craftsmen, explaining disease in terms of natural processes rather than mythological or religious categories, attempted to free the treatment of disease from magic and superstition. From this emerged a type of medical craftsman or physician significantly different from earlier categories of 'physicians'. They were lower in prestige than primitive medicine men, for they were not separated from the rest of society by having access to a mysterious, supernatural, demonic, magical, or divine realm. They did not possess the aura or the regalia of a priest or seer or one divinely-anointed. But they were more than merely craftsmen insofar as they possessed not only technical skill, but theoretical and philosophical knowledge and understanding, which provided both explanations for their techniques and resources for expanding the range of their efficacy. They knew not only the part but the whole *if* they were physicians of Aristotle's second category. In his *Politics*, Aristotle had written that there were three different types of physicians: (1) the ordinary practitioner; (2) the master of the craft; and (3) the man who, as part of his general education, has studied medicine (*Politics* 1282a). The first of these is the type of physician who relies only on experience, and the second is one who combines theory and practice (cf. *Metaphysics* 981a and *Nichomachean Ethics* 1102a).

The separation of medicine from religion did not denote an antagonism to religion. The view of the new medicine was that all diseases are both sacred and natural, thus precluding, in most circumstances, the direct interference of supernatural forces in the natural order. But as the natural order itself was regarded as divine, so every natural event was thought to be divine. Such explanations are entirely consonant with the spirit of an age of which rationalism was the hallmark. In the second half of the fifth century B.C., owing in great part to the widespread influence of the Sophistic movement, for many Greeks naturalistic explanations were displacing the traditional mythological and religious views. This was also a time of considerable lay interest in medical questions, and there were many who belonged to Aristotle's third category of 'physicians' (i.e., who studied medicine as part of a general education). Rationalistic Greeks who turned to rationalistic physicians for help, prophylactic or curative, will have expected to receive rational explanations, advice, and treatment.

Healing shrines continued to thrive side by side with the development of rational medicine. Asclepius came to be the chief healing god of classical antiquity. In the fifth century B.C. his cult began to spread, until there were over 400 temples and shrines dedicated to him in the Mediterranean world. Various types of people were attracted to these: the ill who still attributed their problems to supernatural causes, the poor who could not afford the services of physicians, and those afflicted with diseases that physicians could not cure. There are many attested cases of 'miraculous' healings, that can best be attributed to ancient psychotherapy, aided by various empirical medical and surgical procedures. As time went on, the practices of the temples and shrines of Asclepius became less a matter of irrational or 'miraculous' healings and more a matter of rational therapeutics.

The healing shrines of classical antiquity, then, administered both religious or supernatural medicine and empirical or rational medicine, often as an act of charity. These institutions provided a complement as well as an alternative to rational medicine insofar as they attempted to provide help through means other than rational or empirical. Those administering such care should not be considered physicians in the sense that their predecessors in the ancient Near East could be properly so considered. Supernatural 'medical' care, conducted by priests in any society in which there is no 'scientific' medical system based upon a 'rational' theoretical body of knowledge, is not a 'complementary' or 'alternative' medical resource. It is *the* medical resource and whatever practices within it that appear to us as 'natural' or 'rational' must be seen within their magico-religious context. Whatever medical services exist apart from it are also neither complementary nor alternative, but decidedly subsidiary to it, dealing only with those minor matters that lie within the competence of the craftsman. But when medical craftsmen formulate a theoretical body of knowledge and undertake to treat, on the basis of their explanatory system, conditions that were previously the exclusive territory of those privy to the supernatural (magical or religious or both), and when the cultural climate is conducive to the general approbation of such a development, religious or magico-religious medicine remains no longer *the* medical resource. It becomes a complement to the new medical resource; it is no longer a source of medical care but an *alternative to* medical care. Its practitioners can no longer properly be called physicians, except in a limited sense insofar as they provide 'rational' or empirical medical care as a supplement to their religious or magico-religious services. This change in the history of medicine is due to the influence of Greek rationalistic philosophy. It is a foundational and definitional change, both functional and ideological, in the very basis of medical practice and the determination of who is a physician. It has provided distinctions that generally have remained fundamental in Western civilization. This is not to say that the definition of who is properly called a 'physician' has been, or indeed is, always clear. The area of dispute, however, had changed. For the most part, it has, since the development of Greek rational medicine, not been between magico-religious medicine and rational medicine. Rather the dispute is between those who consider themselves as belonging to the latter category, and it is waged over such matters as competence, technique, theory, professional status, and so forth.

Roman medicine, during the early Republican period (from the fifth to the third century B.C.), was largely a combination of folk medicine and magical incantations. During the late third century B.C., Greek rational medicine first penetrated Rome. It was welcomed by some and severely distrusted by others. Also in the third century B.C., the cult of Asclepius (Aesculapius to the Romans) was introduced into Rome, providing a basically Greek alternative of religious healing to earlier Roman magico-religious practices as well as a quasi-rationalistic and empirical alternative to traditional Roman folk medicine and to incipient Greek rational medicine. Under the Empire (31 B.C. to A.D. 493), particularly in the late Empire, there was a considerable increase in the availability, variety, and popularity of magical and cultic healing practices, mostly of Oriental derivation. This was an era characterized by superstitious attitudes and practices that penetrated into all areas of life including philosophy and medicine, as evidenced by a frequent reliance on astrology and alchemy. Although medical practice became more and more superstitious, in some instances employing magical procedures, and although in doing so it became less rational, it did not return to magico-religious medicine, as typified by the medicine man of primitive societies, or the sacerdotal medicine of ancient Near-Eastern cultures.

The most significant event in the history of Western civilization, given the influence of the religion that bears his messianic title, was the life of Christ. In early Christianity, and indeed up to the present, there has been a tension between the medicine of the soul and the medicine of the body, between the life of faith and a use of, or reliance upon, material or secular means. This tension is particularly seen in the question of whether it is proper for a Christian to resort to secular medicine instead of relying upon spiritual means.[8] Origen, writing in the early third century, held that one seeking to recover from a disease had two alternatives: either to have recourse to the medical art, which he labels as the simple and more ordinary method; or to rise to the higher and better method, namely to seek God's blessing through piety and prayers (*Against Celsus* 8, 60).

These alternatives were not regarded as mutually exclusive in the early church. They could be complementary. Sometimes they were employed in conjunction with one another, sometimes separately. Some Christians combined secular medicine and prayer, while others relied exclusively on prayer. Some turned to prayer only when secular medicine proved ineffective, while others resorted to secular medicine only when prayer seemed not to avail. Some sought divine intervention through faith healing and would have no recourse to physicians. Others would try faith healing only as an act of final desperation. Except for a few sources that show an utter hostility to, and contempt for, secular medicine, even a complete reliance on religious means of healing does not necessarily imply even a disparagement, much less an unequivocal condemnation, of secular medicine.

In the earliest centuries of Christianity, a variety of means of healing appears in the literature – prayer, anointing, laying on of hands, the administration of the sacraments, and exorcism. Beginning in the fourth century, there is a significant change in the accounts of healings: sensationalism and bizarre incidences begin to abound. This change occurred during the same period in which, as was mentioned above, superstitious practices and a general atmosphere conducive to credulity began to typify the ethos of the Mediterranean world. Belief in and practice of magic were widespread. Church fathers (both those who believed in the efficacy of magical means, as most probably did, and those who denied it) and church councils condemned magical practices. Some practices that were essentially pagan, however, were ultimately approved by the church and became a form of 'Christian magic'. Included under this rubric were, e.g., the cross as an all-powerful charm, the name of Jesus as an irresistible spell, and the veneration of saints and martyrs, particularly in connection with relics and shrines. Numerous miracles, particularly of healing and exorcism, were said to have been accomplished by such means.

Christians who chose to avail themselves of secular medicine were advised by some authorities, in addition to avoiding the use of magic, to be circumspect in the use of medical means, realizing that healing comes from God alone and that God sometimes permits illness for reasons that should preclude any solution other than spiritual. A good example of this attitude is found in Basil the Great, who lists six categories of physical afflictions (according to why God permits or sends them), but allows medical treatment in only two of them (*The Long Rule* 55). This can hardly be considered as a disparagement of the medical art in that Basil himself founded a vast charitable institution that included hospitals staffed by priests, physicians, and nurses.

The foundation of Christian hospitals was a reasonable consequence of Christian charity. Christ's commandment to love your neighbor as yourself was a categorical imperative that resulted in, among other things, a strong sense of duty to visit, comfort and care for the sick. In the words of Henry Sigerist, Christianity introduced

> the most revolutionary and decisive change in the attitude of society toward the sick. Christianity came into the world as the religion of healing, as the joyful Gospel of the Redeemer and of Redemption. It addressed itself to the disinherited, to the sick and afflicted, and promised them healing, a restoration both spiritual and physical. . . . It became the duty of the Christian to attend to the sick and the poor of the community. . . . The social position of the sick man thus became fundamentally different from what it had been before. He assumed a preferential position which has been his ever since ([32], pp. 69–70).

Early Christian literature abounds with examples of the compassionate care of the sick, both Christian and pagan. There are few references, however, specifically to Christian physicians from the period when Christians were a persecuted minority. Paul's travelling companion, Luke, is said to have been a physician (Col. 4:14) and we occasionally hear of Christian physicians who were martyred. As the new religion spread, it is certain that more Christians became physicians and more physicians became Christians. The extant sources, however, tell us little about the effect of Christian ideals on actual medical practice. The extent to which Christian physicians conformed to these ideals will probably have been commensurate with their Christian conviction, and occasionally we find mention of physicians like Eustathius, whom Basil praises for his combining a spiritual ministry with his medical practice. Basil writes, "And your profession is the supply vein of health. But, in your case especially, the science is ambidextrous, and you set for yourself higher standards of humanity, not limiting the benefit of your profession to bodily ills, but also contriving the correction of spiritual ills" (*Epistles* 189). It is the effect of his Christian ideals on his "standards of humanity," his concerning himself not only with treating the physical ills of his patients but also ministering to their spiritual ills for which Basil praises Eustathius, not for combining secular and religious *means* of treating physical complaints. From this and other sources there emerges a picture of the ideal Christian physician who combines medical art with spiritual commitment but does not attempt to be a faith healer, discarding rational medicine for supernatural means. A new breed of physicians begins to appear in the literature, fervent Christians whose primary commitment was to Christ and whose secondary commitment was to the medical art, physicians who considered the practice of medicine a vehicle for Christian charity and evangelism. These physicians increasingly, as time went on, tended to be clergy, either priests or monks. There is much evidence from the early Middle Ages for the practice of medical charity by the clergy, especially by monks.[9] Monasteries became the refuge of the persecuted, the poor, and the sick. In many instances such care as could be provided was given by monks who made no claim to extensive medical knowledge and would hardly, even by the standards of that time, be considered as *medici*, physicians. But many examples survive of monks or priests who were regarded as *medici*.

We must be careful here to recognize that clerical or monastic physicians were first and foremost clergy for whom the practice of medicine was an extension of their ministerial role, an act of Christian charity performed for the glory of God and the love of man. It was not sacerdotal medicine as practiced by priests in societies where the supernatural etiology of disease dictated a reliance upon supernatural means of treatment. The medicine practiced by monastic and clerical physicians, although by modern standards riddled with simplistic, erroneous, and sometimes superstitious explanations and procedures, was nevertheless *essentially* rational. The religious functions of these physicians would, at the most, be complementary to their medical efforts and probably directed toward the patients' spiritual rather than physical ills.

A somewhat different situation prevailed at the numerous shrines of pilgrimage that existed throughout and beyond the Middle Ages. Here the emphasis was in great part on the supernatural healing of disease or defect. It was, however, not uncommon to have some medical treatment administered by the clergy at some of these shrines to which the ill and disabled flocked in the hope of a miracle. Given the evidence that many of these shrines possessed excellent medical libraries, it is likely that priests who were also *medici* attempted to treat the ill or relieve their pain.

When the physician is also a priest or monk seeking to succor or cure the ill ostensibly from compassion for them and love for God, the relationship between the physician and the patient is inherently and intrinsically different from that which generally prevailed in the practice of medicine among primitive people, and in ancient Near-Eastern or Greco-Roman cultures. In this respect the observation by Sigerist, quoted above (p. 15), is very significant. Regardless of instances of compassion for the ill evidenced by society in those cultures, Christianity indeed did introduce a marked and essential change in the attitude of society toward the suffering, deprived, and destitute in general and the ill in particular. This new attitude has undergirded the practice of medicine and the position of the patient in Western society ever since, if not always, or even predominantly, in practice, at least as an ideal such that any gross deviation from it is generally viewed as an unworthy or even reprehensible act.

Throughout the Middle Ages there were, in addition to clerical and monastic physicians, secular physicians and medical craftsmen. We know little of this rather amorphous lot during the early Middle Ages. During the later Middle Ages secular physicians began to increase in number and importance and a secular medical profession emerged in a sense that had not existed before and with certain characteristics that have prevailed to the present. We shall speak more of this later. But throughout the Middle Ages, and indeed during the Renaissance and Reformation, two distinct types of medical practitioners existed side by side: the clerical or monastic physicians who did not practice supernatural but rational medicine, and secular physicians who also practiced rational medicine. The first category, for a variety of reasons, gradually decreased in number and thus in significance as the second increased, particularly from the late Middle Ages on. The major difference between the two was in their motivation for practicing medicine. It is undoubtedly true that some secular physicians who pursued medical knowledge and practiced the art were motivated primarily by an intense desire to alleviate the suffering of mankind and, within the Christian context, to glorify God and show love for their fellow man through their medical practice. Also it is certain that some monks and priests practiced medicine in pursuit of filthy lucre, a concern expressed in medieval canon law. But, as a general rule, it is safe to assert that while clerical or monastic physicians practiced the medical art as an extension of their Christian commitment, secular physicians, regardless of how much their conduct was tempered by the expectations of a predominantly Christian society, pursued a medical career for economic reasons. We turn now to a consideration of one of the most prominent and troubling factors in the patient-physician relationship, the economic aspect, in which the physician is the receiver, and the patient is the giver, with its correlate, or opposite, medical philanthropy or charity, the most material evidence of a physician's compassion.

THE STATUS OF THE PHYSICIAN

Rational medicine began in the Greek world as a *techne*, a craft, and it never lost its identity in antiquity.[10] In Homer physicians are grouped with minstrels and builders (*Iliad* II, 514–515, *Odyssey* 17, 382–386). Physicians acquired their skill and knowledge through an apprenticeship in which they learned traditional practices. They might then open their own shop or treat patients in the latters' homes. Many physicians were itinerant and travelled from community to community. There were no medical schools, no examinations, no procedures of licensure. Hence anyone could practice medicine and there was no clear distinction between a physician and a quack, since there was no external authority to guarantee a minimum standard of knowledge or proficiency. Moreover, because medicine was regarded as a craft, physicians enjoyed relatively low status in classical society. Their prestige, even their ability to earn a livelihood, depended on convincing others of their skill and knowledge. Hence they frequently sought out patients, engaged in competition with other physicians to prove themselves better physicians than others, and devised means of advancing their reputation. One such means was prognosis, which was devised, it has been suggested, as a means of gaining the confidence of patients, increasing physicians' prestige, and protecting them against blame for failure to cure a disease ([15], pp. 65–85, esp. 69–70).

The development of a medical science in the fifth century B.C. did not lead to the disappearance of the unlearned practitioner. In fact, throughout Greco-Roman antiquity the 'leech' or ordinary practitioner remained the most common type of physician. The average physician remained for a thousand years a craftsman, unlicensed, and trained, if at all, as an apprentice. The number of physicians who undertook a 'scientific' approach to medicine must always have been relatively small. Concentrated, most likely, in large cities, where possibilities of research and informal collaboration were possible, a few came in time to recognize the value of physiology and anatomy, and sometimes performed experiments and dissections. Yet even in research, medicine was heavily influenced by philosophy. It was the philosophers who posed the basic questions of method, who developed the epistemological systems, and who provided the general theoretical orientation for the educated physician ([15], pp. 352–356). Unlike modern medicine, which is empirical, 'scientific' medicine in classical antiquity was saddled with a great deal of philosophical speculation. Naturally, different physicians followed different philosophical views, and it was from these different views that the medical sects emerged. Once theory entered medicine, 'scientific' physicians aligned themselves with one of the sects, some merely to give the appearance of learning for the sake of prestige, others because of a sincere attachment to the doctrines of the sect ([34], pp. 137–153, esp. 150). But Galen claims that in his own day the majority of physicians were ignorant not only of general studies, but of anatomy and prognostics as well.

The position of the medical doctor in classical society was an insecure one, and many physicians felt this insecurity keenly. In part this was due to the lack of any professional standards for medicine and the necessity of competing against all kinds of healers from magicians to herbalists. Length of training or study at a famous medical center or under a well-known physician might provide credentials, but what really mattered to his patients was his competence in the medical art. Hence all physicians were supremely concerned for their reputation, a subject which is addressed frequently in the Hippocratic Corpus. Physicians must gain the confidence of their patients since patients are concerned whether physicians are competent to treat them. Prognosis is a useful tool in impressing the patient with the physician's knowledge and in absolving the physician of blame in the event of the patient's death. The uncertain position of the doctor was also due to his being regarded as a mere craftsman. The modern distinction between the arts and the crafts did not exist in the classical world. The Greek word *techne* and the Latin *ars* were used for both as well as for what we should call a science. The physician, as a craftsman, had to endure the social stigma attached to members of the 'banausic' trades. This stigma arose from the fact that craftsmen engaged in their trades in order to earn a living. In a well-known passage in the *Politics*, Aristotle distinguishes between the liberal and illiberal arts by describing those as illiberal that weaken the body and earn wages. The distinction also depended on the object in view: "If one follows it for the sake of oneself or one's friends, or on moral grounds, it is not illiberal, but the man who follows the same pursuit because of other people would often appear to be acting in a menial and servile manner" (*Politics* 1337b, trans. Rackham). It is not work *per se* that is condemned here, for while the Greeks did not exalt the value of labor and attribute to it moral value, they did not condemn it as degrading. What was degrading was to work for another person and to be dependent on that person. To work for a wage imposed a limitation on one's personal freedom. For a free person to work for another involved an element of servitude. To be in a state of economic dependence on someone else and at the mercy of another's approval violated the ideal of personal self-sufficiency: in that sense it approximated slavery ([28], pp. 25–30). It is sometimes argued that this view represents aristocratic prejudice and was not the view of the ordinary Greek. But it was the upper classes who to a large degree moulded public opinion and we know from the speeches of the fourth-century orators that similar views were held by Athenians who were not members of the upper class.[11]

As craftsmen, physicians charged fees for their services. There is an interesting discussion about fees in the pseudo-Hippocratic *Precepts* (4–7), in which it is recommended that one not be anxious about fees lest the issue affect the health of the patient. The physician is encouraged to place reputation above profit and to offer services without charge in return for a previous favor or if the patient is without means. Since physicians earned a living by the practice of medicine they were subject to the social stigma of the artisan. On the other hand it was recognized that they should have the interest of their patients as their primary concern, as the *Precepts* advises. The matter is discussed at length in Plato's *Republic* (340C–347A), where the question is debated whether the physician as a physician is a healer or an earner of fees. Socrates argues that the physician does not seek the advantage of medicine or the physician but only the advantage of the patient. Medicine, like every art, is concerned for the good of its object. The earning of fees from medicine is a subsidiary art that is attached to medicine or any craft to provide the motivation for undertaking that craft. But there need be no conflict between the motivation of the physician (earning a living) and the aim of the art (healing the sick). In spite of the fact that Greek physicians belonged to the lower strata of society and that they were often men who travelled from place to place and therefore not members of a community, some physicians gained good reputations. Plato treats physicians with respect in his dialogues: Eryximachus, for example, is depicted as a social equal with the other guests in the *Symposium* ([15], pp. 153–171). Some became public physicians of Greek cities; some in Hellenistic times became court physicians; still others gained international reputations and made enormous sums of money. Complaints of physicians' greed and high fees are made by writers throughout the classical period.

In contrast to the Greeks, the Romans depended largely upon freedmen and slaves for medical treatment. As late as the first century B.C. no Roman citizen practiced medicine and Pliny says that Romans seldom entered the profession. Most physicians in Rome were Greek or Oriental. The Romans seem to have preferred foreign physicians and Greek was the language of medicine. Julius Caesar gave Roman citizenship to all permanent residents of Rome who practiced medicine. In 23 B.C. Augustus was cured of a serious illness by his physician, Antonius Musa, who was a freedman, and the Emperor as a result granted to him and to all physicians in the future immunity from civic obligations. Physicians benefitted from immunities that were extended by later emperors, which made the profession an attractive one and may well have drawn many to medicine. These immunities from public burdens and exemptions from taxation generally did not extend to physicians in the lower strata of the social scale, however. A significant feature of Roman medicine was the number of slaves engaged in medical practice. Many wealthy families had their own household physician. Some large aristocratic households maintained both male and female medical slaves. Apparently some physicians who taught medicine readily accepted slaves as students because of money that they could earn thereby, a practice that is condemned by Galen. In 93–94 A.D. the Emperor Domitian issued a rescript forbidding this practice, which was apparently causing some concern ([37], pp. 114–115).

The low status of most physicians posed a problem for members of the Roman upper class who found it necessary to place themselves under the care of a physician who was a social inferior. It was a reversal of the natural order for free persons to obey slaves, even if a slave was one's physician; yet it was essential for patients to accept the authority of their physician. Undoubtedly social factors entered into the treatment of patients by their physicians in Rome, just as Plato points out that a physician found it necessary to treat free persons in a different way from slaves: "He gives no prescription until he has somehow gained the invalid's consent; then, coaxing him into continued cooperation, he tries to complete his restoration to health" (*Laws* 720, trans. Saunders). There were, of course, some physicians who overcame the limitation of their background: slaves who gained their freedom; those who came to hold the office of public physician in cities throughout the Roman Empire and who gained immunities from civic burdens and taxation; and a few who became court physicians. Some became teachers and founded medical sects. A few became wealthy. Quintus Petronius had an income of over half a million sesterces a year. Crinas of Massilia, who practiced during the reign of Nero, left at his death ten million sesterces, after having spent an equivalent sum in building the walls of his native city. Alcon, who was a surgeon during the reign of Claudius, was fined ten million sesterces and banished to Gaul; yet upon his return he gained that amount again after a few years' practice. Pliny the Elder, who is notorious for his hostility to physicians, writes that there is no art that is more profitable than medicine (*Natural History* 29.2).

Yet wealth could never completely overcome the prejudice of the Roman upper class against physicians, who always remained craftsmen. It is true that attempts were made to elevate medicine by identifying it with the 'liberal' arts. The Greeks, and later the Romans, came to divide the arts into 'liberal' and 'vulgar' or 'sordid' (*artes liberales* and *artes vulgares* or *sordidae*). The vulgar arts are those that require physical effort, while the liberal arts have an intellectual content that make them superior. In a well-known passage in the *De Officiis*, Cicero classifies various means of livelihood according to whether they are vulgar or liberal arts. He mentions three that are distinguished by their intellectual content and their service to society, namely, medicine, architecture, and teaching (*De Officiis* 1.150). He commends them as honorable for those to whose social standing they are suited. But there was no agreement about whether medicine was truly a liberal art and it seems to have hovered on the fringe of the classical canon of the truly liberal arts. Although there was much discussion regarding the place of medicine it seems not to have gained full acceptance in the classical world as a liberal art and therefore never strictly became one of the 'honorable' professions.

The standing of medicine as a profession had a bearing on the method of remuneration for services. Part of the disdain for craftsmen was that they worked to order for fees. It was considered dishonorable for a member of the 'professions' or liberal arts to request payment; they might accept an honorarium, but not request payment. Hence the 'professional' did not share in the stigma of dependence that a craftsman did. The status of physicians as regards remuneration has been much debated. Were they governed by hire of services or by mandate, i.e., an ostensibly gratuitous service that brought with it an honorarium? Presumably physicians who were of freedman or servile status always accepted fees for their services. More prominent physicians were probably governed by mandate: Galen, for example, says that he never requested payment, yet he accepted it, no doubt as an honorarium for his 'gratuitous' services. But difficulties remain and not every case can be explained in a fully satisfactory manner, in part because of changes in the social status of the profession.[12]

In spite of the reputation that many physicians enjoyed in the Roman Empire the medical art never fully escaped its standing as a craft. The result was a sense of insecurity that necessitated the frequent defense of medicine as a genuine art. Hippocratic physicians were often required to reply to their detractors' charges of ignorance, incompetence, or lack of proper knowledge or technique. At issue was the authority of physicians and their ability to inspire the confidence of their patients. Indeed, throughout antiquity this was the chief problem facing physicians in their relationship with their patients. Galen, for example, stresses that the physician must convince the patient of the benefits to be derived from following the doctor's orders. He even suggests that the physician should so endeavor to win his patient's respect that the latter would admire him like a god. Although the physician in Roman times no longer had to defend the art of medicine itself, the conflicting claims and mutual hostility of the medical sects revealed that the basis for medical practice still remained an issue. And although medicine claimed to have a body of knowledge of its own and the means to train its practitioners, physicians were still, in the words of H. I. Marrou,

> dogged by an inferiority complex. From Hippocrates to Galen they go on saying: 'The physician is a philosopher as well'. They had no desire to remain walled up within their own particular culture; they longed to join in with the general culture on a genuinely human level; and for this they did not rely on their technical training but, as can be seen in Roman times in the case of Galen, they tried to be educated men like all the others, men who knew their classics, men who could speak like rhetors and argue like philosophers ([26], p. 303).

Galen's views on the matter are set forth in a short treatise entitled *That the Best Physician is Also a Philosopher*, which argues the necessity of a physician's being acquainted with other areas of knowledge besides medicine. Galen believed that philosophy provided the basis for medical research, treatment, and ethics. One need not doubt the genuine sincerity of Galen to recognize in his attempt to elevate the status of medicine by grounding it in philosophy a perhaps unconscious desire to establish more securely medicine's authority.

As we have seen, there were in the early Middle Ages two distinct groups of medical practitioners: (1) clerical and monastic physicians and (2) secular medical craftsmen. Of the latter there was a fairly diverse variety of which we have relatively little knowledge. Their relations with their patients, their relative status, and their fee policies are matters about which one may speculate, but about which insufficient information exists to provide the historian with anything more than occasional anecdotes. Clerical and monastic physicians, if they were practicing medicine with the motivation proper for those whose lives were ostensibly devoted to love of God and service to their fellow man, should have acted as compassionate vessels of charity to their patients. During the high and late Middle Ages we begin to encounter fewer clerical and monastic physicians, while secular practitioners appear more and more in the sources. There is still a good deal of diversity of types among these practitioners, but particular types become more easily identifiable. Although generalizations are not safe in this matter, it is reasonable, although perhaps overly simplified, to say that a major division begins to appear by the late Middle Ages between those practitioners who have been trained in the burgeoning medical schools of the universities and those who have learned their trade through apprenticeship. With the advent of medical licensure regulations and craft guilds, physicians (*medici* or *physici*) were generally university trained, while manual operators, i.e., surgeons, were trained within their craft guilds. Although there are many significant exceptions, particularly in southern Europe, nevertheless, in great part during the late Middle Ages and increasingly during the Renaissance a division between medicine and surgery deepened and the concern with status, especially by physicians, caused them to recoil from the more manual and physically unpleasant aspects of their work. In some quarters medical practice was so affected by academic pedantry of the most abstract and theoretical nature, that physicians would not even deign to see patients directly but, working through a runner, would examine a urine specimen, diagnose the illness, and prescribe treatment. Surgeons, who, in many locations, were not permitted to treat conditions other than those that yield to surgical intervention, often engaged in battles with medical faculties to acquire university status for their discipline.

There was, at times, a void in medical and surgical care, a void filled both by various 'quacks', particularly women, unlicensed, of course, who were willing to treat the medical ills of the afflicted, and by barbers, licensed to administer minor surgical care, who were willing to treat cases that surgeons were increasingly coming to regard as beneath their dignity. Many cases came to court in which physicians' organizations charged unlicensed practitioners with illegally treating the ill. Testimony is introduced by the defense, usually of former patients who praise the compassion, tenderness, and availability of the defendant and the efficacy of his (or, many times, her) treatment is contrasted to the relative unavailability of the lofty and arrogant university-trained physicians, who usually were members of the university faculty as well, physicians who, when they would deign to treat, even through a runner, were, according to the witnesses, singularly unsuccessful in their efforts. Without fail, regardless of how effective their treatment and how much they were respected and loved by those whom they had helped, these unlicensed practitioners were found guilty and forbidden to practice, or punished if it was a second offense, on the ground, tacitly, that it was competence as gauged by training and licensure and not by success in treatment that was requisite to the practice of medicine. Such examples as are available from Paris [23] may represent an extreme. Even if the situation there was extreme, however, it certainly was not a complete anomaly for Europe during the late Middle Ages.

There were, of course, physicians and surgeons during these centuries who were conscientious and competent, given contemporary standards of competence, those who, in earning their livelihood by medical or surgical practice, endeavored to exercise their art or craft ethically and in a manner fully consistent with the expectations of a society that in ideals, if not in reality, was permeated by Christian principles. It was in the realm of fees and charity that conscientious physicians of the late Middle Ages and Renaissance experienced their greatest frustrations. The writings of Henri de Mondeville, a surgeon who lived in the late thirteenth and early fourteenth centuries, nicely illustrate these frustrations. He suggests that the surgeon not "have too much faith in appearances. Rich people have a habit of appearing before him in old clothes, or if they do happen to be well dressed, make up all sorts of excuses for demanding lower fees." These patients

> claim that charity is a flower when they find someone else who will help the poor, and thus think that a surgeon should help the unfortunate; they, however, would never be bound by this rule. . . . I tell these people, then pay me for yourself and for three paupers and I will help them as well as you. But they never answer me, and I have never found a person in any position, whether clerk or layman, who was rich enough, or honest enough, to pay what he had promised until I made him do so ([20], p. 156).

Accordingly, "the surgeon ought to charge the rich man as much as possible and get all he can out of them, provided that he does all that he can to cure the poor" ([36], pp. 356 f.). The conflict between the physician's economic needs and the patient's desire for Christian charity is a frequently-encountered theme, one not seen in pagan societies.

THE PHYSICIAN AND PHILANTHROPY

To what extent was there a philanthropic or humanitarian impulse in the relationship of physicians to their patients in Greco-Roman medicine? Were

physicians in classical times motivated by a love of humanity that added to their merely professional interest in their patients a desire to extend to them personal concern and compassionate care? In order to answer these questions we must first examine the place of philanthropy in the classical world.[13] The Greek word *philanthropia*, from which our English word is derived, had a variety of meanings. In its root sense it means 'love of mankind' but, as often used, it means little more than 'kindly or friendly' and it is frequently used in this sense in the Hippocratic Corpus. Thus in a well-known passage in *Precepts* (6) the statement is made that "where there is love of man (*philanthropia*) there is also love of the art (*philotechnia*)." Although this passage has often been cited as a proof-text for the existence of a philanthropic motive in Greek medicine,[14] its actual meaning is very different. The physician here is simply being encouraged to show kindness, in return for which a patient can be expected to exhibit a love of the physician's art. This trust, as it were, in the physician's skill brings in turn contentment with the physician's goodness and aids in the curative process. This advice to physicians to display a kindly attitude towards their patients is not an exhortation to philanthropy. It appears, in fact, in a discussion dealing with fees. The 'kindliness' that is urged on physicians in the Hippocratic Corpus is intended to be merely an aspect of the physician's professional deportment.

The restricted sense in which *philanthropia* is used in Greek reflects the limitation in classical society generally of ideas in philanthropy. There was little in the Greco-Roman world that resembled private charity or personal concern for the needy. Pity was not regarded as a proper motive for assisting those in need except members of one's own class, particularly members of the upper class, who had experienced a reversal of fortune. There was the recognition that one might aid others on occasion because of feelings of common humanity, but this benevolence was limited and largely given with the hope that it might be reciprocated if one someday needed similar assistance. There was, in short, no moral or religious basis for charity or what could be called humanitarian aid. There is no evidence in the Hippocratic Corpus that physicians are expected to make humanitarian concern a part of their approach to medical treatment.

Yet civic philanthropy played an important role in the classical world. Philanthropy in Greece and Rome generally took the form of civic benefactions rather than personal charity. Its motivation was not pity but a desire for public recognition based on the assumption that the conferring of a benefit should entail repayment. Reciprocity was considered to be basic in giving any gift or conferring any benefit. One way in which a public benefactor could be repaid was by means of expression of honor by a community or a portion of its members. It was a widespread practice in the classical world for distinguished or wealthy members of a community to bestow a public benefaction for which they expected public recognition. This recognition was commonly given in the form of an honorary inscription that gratified the *philotimia* ('love of honor') of the giver by increasing his prestige or standing in his community. Civic philanthropy was practiced by physicians, for whom we have a large number of honorary inscriptions set up by grateful cities in Hellenistic Greece in the third and second centuries B.C. These inscriptions record their tireless service to their cities, their dedication to their profession, their treatment of all members of the community, whether rich or poor, citizen or foreigner, their remission of fees, and their devotion to duty during epidemics. There is little in these inscriptions that is personally descriptive of the physicians honored. The language is formulaic and similar to that used in public decrees that honor members of other professions. As members of the community, physicians were expected to offer their services to all without distinction; accordingly, their philanthropy was civic and not personal. In return for distinguished public service they were honored by the community as a whole. A desire for this kind of honor furnished an important motivation for many classical physicians. To the Greeks it was expected that a good person would seek honor. Because a good reputation was important to physicians in a society that lacked any form of official certification, public recognition increased the confidence of those who might seek treatment from them. The lack of a philanthropic impulse in physicians will not have appeared to be a deficiency for most potential patients. Pre-Hellenistic Greeks showed little concern with the motivation of a physician; what really mattered was competence in the art and public honor gave implicit recognition of a physician's ability.

After the fourth century B.C. the concept of philanthropy began to undergo a change and it came to express cosmopolitan and humanitarian ideas that were lacking earlier. The philosophical sects of the Stoics and Cynics emphasized the universal kinship of mankind and their views influenced both philosophical and popular ethics in Hellenistic Greece and in the Roman Empire. Aulus Gellius (second century A.D.) says that the Latin word *humanitas* was commonly used in his own day with the same meaning as the Greek word *philanthropia*, which indicated "a kind of friendly spirit and good feeling towards all men without distinction" (*Attic Nights* 13, 17, 1). One finds this concept of philanthropy, which was predicated on a belief in the brotherhood of all humanity, in the writings of Galen, who was influenced by Stoic beliefs. His concern with the philanthropic aspects of the medical art is set forth in his work *That the Best Physician is Also a Philosopher*. For Galen medicine was "an especially philanthropic art" because it alleviated the sufferings of humanity and advanced and disseminated medical knowledge for the sake of all mankind, present and future. The model of the philanthropic physician for Galen was Hippocrates, who epitomized these virtues. Another medical writer who reveals an even deeper influence of Stoicism on philanthropy in the practice of medicine is Scribonius Largus, a Roman physician who lived in the first century of the Christian era. His views are set forth in a short essay entitled *Professio medici*, which is prefaced to his treatise *On Remedies*. While for Galen a compassionate attitude on the part of a physician is a desirable quality, it is not (as is competence) an essential feature of the practice of medicine. For Scribonius, on the other hand, the physician should have a compassionate heart that is characterized by *humanitas*, which he considered essential to the profession of medicine.

To what extent were the views of Galen and Scribonius Largus reflected in actual medical practice? Did they represent only the thinking of a small number of educated physicians, who were affected by the prevailing attitudes of philosophical ethics or did these ideals influence the relationship of physicians to their patients on a more general level? While we have no way of knowing with absolute certainty, it is not unlikely that many physicians came to believe that the patient should be treated not just as the object of the physician's skill but as an individual to whom the physician owed succor and compassion that was due all without distinction. There is evidence that the Stoic doctrines of cosmopolitanism and human brotherhood influenced Roman attitudes towards slavery; and it is hard to believe that they did not similarly influence the attitudes of many physicians to the extent of adding another dimension to the physician-patient relationship.

As we have seen, the advent of Christianity introduced a marked change in attitudes towards the sick and the afflicted. This change reflected the influence of the Christian emphasis on personal compassion and charity that was foreign to classical medical ethics, at least before it appeared on a limited scale in some Hellenistic and Roman authors. The Christian understanding, however, was rooted not in classical concepts of *philanthropia* but in *agape*, a word that was taken over from classical usage and given a specifically Christian meaning. *Agape* denoted the love of God as revealed in Christ, love that was unbounded, spontaneous, sacrificial, and not dependent on the goodness of its object. Christian love imitated God's love and provided the motive for Christian behavior. It was an active principle that was expected to be worked out in acts of personal charity and philanthropy. These acts of charity included relieving physical affliction (see Matthew 22:36–40 and Luke 10:25–37); and they were incumbent upon all Christians, as they were expected to demonstrate their devotion to their Lord by deeds of kindness and sacrificial care of those in need. The early Christians took seriously their duty to help the sick. Deacons and deaconesses had a special obligation in this regard and women especially were considered to be able to devote their time to visiting the sick. Instances are recorded of heroic service of Christians on behalf of the sick in times of pestilence, and even Roman pagans remarked on the contrast between the concern shown by the Christians for those in need and the indifference of the majority of the population.

In 313 A.D. the Emperor Constantine made Christianity a fully legal religion, thereby ending two and a half centuries of persecution. Christianity grew rapidly and Christian values gradually came to replace pagan ones in the course of the fourth century. We have already noted that hospitals began to be established by Christians throughout the Roman Empire. Orphanages and homes for the poor and the aged were founded as well. These hospitals (*xenodochia*), the first general institutions of that type in the world were recognized as specifically Christian foundations; they were occasionally established by private endowment, but usually by bishops or monasteries. With the rise of the monastic movement in the eastern Roman Empire and the spread of monastic communities to the West in the fourth and fifth centuries, medical charity received much emphasis as an aspect of the charitable work of monks.

Christian teaching, with its belief in God's love of all humanity and its exhortation to demonstrate the love of Christ to those in need, was responsible for a radically new element in the motivation for the practice of medicine and the care of the sick. In treating the ill the physician was to be moved by his love of God and his desire to serve Him. Accordingly, Augustine says that the ideal physician is motivated by charity and is willing to treat the poor without any thought of remuneration. It is impossible, of course, to say how extensively the Christian ideal was practiced by Christian physicians. Numerous examples can be cited from early Christian literature of physicians, both lay and clerical, who are praised for their devotion to Christian ideals in their medical practice. One presumes, however, that, human nature being what it is, many physicians were little influenced by the demands of Christianity on their profession. Nevertheless, whether honored sincerely or in the breach, a higher and deeper conception of the patient-physician realtionship resulted from the teachings of Christianity that was to dominate medical ethics for over a millennium.

ATTITUDES TOWARD PHYSICIANS

Attitudes toward physicians have varied considerably throughout the history of mankind. Will Durant made a perspicacious observation when he said that "in all civilized lands and times physicians have rivaled women for the distinction of being the most desirable and satirized of mankind" ([13], p. 531). The validity of this assertion can be demonstrated by anyone who pursues the numerous references under 'physician' in Stith Thompson's *Motif-Index of Folk-Literature* [35]. The motifs that have become attached to the healer in diverse cultures are so constant that they appear to transcend time, place and ethnocultural barriers.[15] The attitudes displayed toward physicians vary, ranging from adoration to contempt, with a large area in between where the foibles to which physicians are most susceptible are treated, sometimes with sarcasm, sometimes with good-natured humor. Although these are usually recounted by the fabler or the raconteur, they reveal, perhaps in a somewhat perverse and distorted way, some of the tensions inherent in the patient-physician relationship.

Before discussing these motifs, however, an important distinction needs to be made. The greater the aura that surrounds 'physicians', the less the tensions revealed by these motifs seem to be attached to them. When 'physicians' are medicine men, the most powerful and venerated persons in the community, whose ability to harness magical forces and to mediate or manipulate spiritual (divine or demonic) beings sets them far above the rest of the community, or when 'physicians' are priests, practicing magico-religious medicine, set apart from others by a special relationship with the supernatural and a knowledge of magic, they enjoy a position of respect and authority far above that of the medical craftsman or scientist of any period or culture. The near splendor that surrounds them lends first to their office and then to their person a level of esteem that gives to their relationship to their patients a quality more sacerdotal than would, except under the very rarest of circumstances, exist in the relationship of the medical craftsmen-physicians with their patients. But it is especially when physicians practice rational medicine and are separated from the rest of society by the quantity, as it were, not the intrinsic quality, of their specialized knowledge, that the tensions that underlie their relationship to their patients surface.

The helplessness of the patient in the hands of the physician is a relatively common theme in literature. The knowledge that physicians possess is believed to give them a greater ability not only to help but also to harm. The potential for harming, if physicians are competent, lies in their capacity for using their knowledge to an evil end, e.g., as a poisoner. Also, physicians, because of their access both to the patient and to the patient's home, if they are unscrupulous, may avail themselves of opportunities for dishonest or immoral activities.

The incompetence of some physicians is frequently lampooned in classical, medieval, and Renaissance literature. The physician who became an undertaker and now puts his patients to bed in his old effective way, or the physician who was an eye-specialist and so accomplished at destroying his patients' sight that he became a gladiator so that he could further perfect his skill, are examples of types regularly encountered. The unhealthy physician who should heed the proverbial advice, 'physician, heal yourself', is held up as one to whom a sick man should know better than to have recourse. The negligent physician is often derided. The pompous or ostentatious physician is a butt of ridicule, especially one who is also quite incompetent in or ignorant of his art. The dour, harsh, tactless, or uncompassionate physician, although not as severely castigated as the evil, negligent, or incompetent physician, comes in for criticism. The pathetico-humorous nature of the pawing hands and peering eyes of medical apprentices or students being instructed by their physician-teacher on the patient-subject-victim, is also a motif in literature, as is the inquisitive physician who, with knife or drugs, is driven to try various novelties to satiate a perverse curiosity or expand medical knowledge at the expense of his patients. It is a frequently-encountered sentiment that 'only physicians can commit murder with impunity' or that 'physicians bury their mistakes'. In all these instances, whether exploited for their humorous potential, albeit with tragic overtones, or expressed with anger, frustration, and despair, it is the helpless, vulnerable patient who is the victim. And in the background there is the ubiquitous voice crying out, "You don't need these physicians, after all, since you can treat yourself." In the opinion of such a nay-sayer, physicians are able to exploit the naive public in great part only because the latter are ignorant of the salubrious remedies that anyone can glean from the fields or buy from the local herbalist. Here again patients – all patients, indeed the public at large – are the victims of a profession that is wilfully, stubbornly, and obstinately ignorant of the simplicity with which health can be achieved and maintained.

It is not, of course, merely a negative picture of the physician that emerges from the literature. Throughout classical antiquity, the Middle Ages, and the Renaissance the physician is often depicted as an ideal and the ideal physician appears as a frequent type in the sources. The physician as an ideal is seen in the highly favorable connotation of the word 'physician' when used metaphorically. Unless modified by a pejorative adjective, it denotes a 'compassionate, objective, unselfish person dedicated to his responsibilities'. Thus the good ruler, legislator, or statesman is sometimes called the physician of the state and, when particularly harsh measures are necessary for its good, is called the surgeon of the state. The underlying sentiment is that the statesman should be to the state what the physician is to a patient. The metaphor is also applied, in classical antiquity, to the relationship of philosophers to those whom they seek to improve by their instruction. The philosopher is seen as a physician of the soul. This idea is seized upon by early Christian authors who refer to Christ as 'The True Physician', 'The Great Physician', or 'The Only Physician', and describe the clergy in general as physicians of the soul. Because such high associations become attached to the word 'physician' when used metaphorically, we can assume that the physician as an ideal type was recognized, regardless of how short of the ideal many physicians fell.

Similarly the concept of the ideal physician appears in both medical and non-medical literature throughout classical antiquity, the Middle Ages, and the Renaissance. While, as we have seen above, philanthropy has not always been regarded as an essential feature of the ideal physician, competence and (to a degree) compassion have been. The conspicuous absence of the latter has usually been held to violate an inherent, although somewhat nebulous and elusive, *conditio sine qua non* of the medical ideal. It was the influence of Stoicism and, much more significantly, the basic principles of Christian *agape*, that tempered the medical ideal with selflessness, sympathy, and, in the fullest sense of the word, caring, that manifested itself without regard to the class or status of the one needing medical care. In the deontological literature of the early Middle Ages, we see a blending of the highly practical features of classical medical etiquette with Christian compassion and charity, that created a lofty ideal of medical practice.

Whenever an ideal exists, minor and major deviations from it are bound to occur. There is probably no relationship in which failure to realize the ideal excites a greater degree of resentment and hostility than in the patient-physician relationship, especially since the ideal represents ordinary expectations of patients in particular. At one end of the spectrum are people who respect or venerate physicians in general and their physician in particular. They also experience the underlying tension in the relationship that is generated by the patient's feeling of helplessness, of being at the physician's mercy. They have, nevertheless, a high level of trust in their physician and often attach an unrealistic aura to medical practitioners, believing that they have the ability to accomplish much more than it is reasonable to expect. When physicians fail, these people, because of their unrealistic expectations, may suspect them of negligence, of evil intent, or at least of not really caring and trying, reasoning that if they had wanted, they could have done more. Or they may view them as incompetent because of an equally unrealistic degree of expertise that they have imagined them to possess. At the other end of the spectrum are those who hold the medical profession in distrust or contempt. They believe that physicians, for the most part, are not only incompetent, but greedy for honor and wealth as well, unprincipled persons who victimize the vulnerable, defraud the helpless, and deceive the trusting. Most people, of course, fall somewhere between these two extreme attitudes toward medical practitioners.

Throughout the history of mankind, regardless of the time or place, there have always been some medical practitioners who were incompetent or negligent, practitioners who richly deserved to be held responsible for their misdeeds. There have also always been conscientious and, for the time, competent physicians who have been quite unjustly blamed for their lack of success in particular cases. It is probably safe to say that the attitude of the individual patient, whether to the medical profession as a whole or to a specific physician, has always been a determining factor in deciding whether or not to take action against a physician, especially as the action is thought to center upon the physician's competence, intent, diligence, or negligence. It is also safe to say that there has usually been an awareness of the potential for harm in rational (as distinct from magico-religious) medical practice. Generally, those who assume a healing role, other than magico-religious, are seen as also assuming the responsibility of possessing and maintaining a certain level of competence and knowledge and exercising a reasonable level of diligence. The level, of course, varies from culture to culture and from time to time. Throughout the course of Western history, the degree of legal sophistication evident in attempts to define and regulate these responsibilities has also varied considerably. In some instances there have been attempts made to define professional responsibilities along lines designed to protect the rights of the patient by providing mechanisms whereby they may seek redress when their rights are thought to have been violated. These attempts at legal definition are usually subsumed under laws of contract or tort and redress is sought by aggrieved patients or their relatives. Different in their intent and motivation are laws not designed to define and protect the rights of the individual patient, but promulgated and enforced at least ostensibly for the common good, such as setting standards, prohibiting certain types of procedures, and prescribing others. Initiators of action against violators of these laws are generally public officials rather than individuals harmed by negligent or incompetent physicians. The more detailed and confining the regulation of medical practice in this second category has been, the fewer mechanisms there have been available to enable aggrieved patients to seek redress under the first category of actions.

THE ETHICS OF THE PHYSICIAN-PATIENT RELATIONSHIP

Historically, attitudes toward physicians run the gamut from idolizing, adoring, respecting, questioning, distrusting, disliking, to despising. These attitudes are, in great part, governed by something much more personal than questions of competence and diligence, namely the patient's desire to have a special place within the physician's priorities. Is the physician concerned with me? Does he really care? Does he see me simply as an object in need of repair? As a source of revenue? Would he sacrifice my well-being, or risk my recovery, by experimenting on me? Am I in some way perhaps special to my physician? The philosopher Seneca, in the first century A.D., writes that if his physician treats him as any other patient, he owes him only the fee for his services. But if his physician gives him more attention than is professionally necessary, even if this is at the expense of his other patients, then "such a man has placed me under obligation, not as a physician, but as a friend" (*On Benefits* 6, 16). There is an underlying selfishness in this attitude, a selfishness that is not infrequently seen in the hopes, if not expectations, of some patients encountered in the literature of our period.

"You're telling me the truth, aren't you, doctor?" "If it's really bad, I don't want to know." "If Dad isn't going to make it, he'll want to know. But please don't tell him." What to tell the patient has been, throughout the history of medicine, a problem, both from the physician's and the patient's (or his relatives') perspective. In the Hippocratic Corpus it is suggested that the physician reveal nothing of the patient's future or present condition, for many patients, on account of this, have taken a turn for the worse (*Decorum* 16). Other medical authors suggest that the physician should tell seriously ill patients of their condition so that they can put their affairs in order. Comments by both Cicero (*De divinatione* 2, 22, 55) and Pliny the Younger (*Epistles* 1, 22) suggest that it was not at all uncommon for Roman physicians to withhold unfavorable information from patients. That some physicians informed at least their terminal patients is evident from Galen's indignation at the callousness with which this was sometimes done. However, when the destiny of the patient's soul may be affected by the physician's decision, the matter becomes much more crucial and voices other than the physician's or the patient's (and his relatives') are raised. During the late Middle Ages and Renaissance moral theologians (summists) wrestled with this question. The opinion of these summists is generally as follows. Informing terminally-ill patients of their hopeless condition, or seriously-ill patients that they might well die, can do grave harm to their physical state. Similarly, failing to tell such patients of their condition can do grave harm to them (1) spiritually, since, if they know their physical state, they would see to making final provision for their soul; and (2) materially, since patients would wish to ensure that arrangements be made for the disposition of their possessions, thus lessening the potential for strife among their heirs. The physician should consider informing the patient through a third party, e.g., a relative of the patient or a priest, or suggesting that a third party at least inquire concerning the patient's spiritual and material state. Of this the summists were certain: the onus was upon physicians to ensure that they would not allow a patient to die without regard for these matters. A physician who fails to make provisions for this has sinned mortally.

"You're going to stick with me to the end, aren't you doctor?" This has been a concern of many patients, at least in modern times. It has however, not been, by any means, a constant feature in the history of medicine, but reflects a concern generated by a significant change in the role of physicians and in the expectations of their patients in Western history. In Egypt physicians usually made a prognosis before undertaking treatment. If the prognosis was favorable or uncertain, they would take on the case; if unfavorable, they usually would not. In Mesopotamia the *āšipu*, in practicing magico-religious medicine, did not take on hopeless cases, nor did he hesitate to withdraw from a case once it was evident that there was no hope of a favorable outcome. The *asû*, the medical craftsman, however, appears to have stayed with his patients to the end. A partial explanation of this difference is that the former usually dealt with chronic conditions, while the latter generally treated acute illnesses and injuries. In Persia, because of the peculiar nature of Zoroastrian dualism and medicine's place within it, physicians apparently did not refuse to take on hopeless cases and stayed with their patients to the end ([4], p. 881).

There was a saying in classical antiquity that the best physicians did not cause their patients to linger on, but buried them quickly. Plato is well known for opposing any efforts to prolong the lives of the ill or debilitated who had no chance of regaining their health. Health was considered both a virtue and an indicator of virtue by Greeks of the fifth century B.C. and later. Without health all else was without value. The author of a short treatise in the Hippocratic Corpus defines medicine as having three roles: (1) doing away with the sufferings of the sick, (2) lessening the violence of their disease, (3) and refusing to treat those who are overmastered by their diseases, realizing that in such cases medicine is powerless (*The Art* 3). The third role reflects the prevailing sentiment among Greek and Roman physicians. The thought seems to have been, at least in part, that the patient's interests would not be served if his life, or dying process, were simply prolonged. But even more important for the ancient physician was the damage that terminal cases had on a physician's reputation. In the absence of licensure, a physician's reputation was his only credential. Achieving and maintaining a good reputation was a delicate business. Charlatans were criticized for refusing to take on dangerous or uncertain cases or, if they did so, for exploiting the patient's hopeless condition for pecuniary gain. But the conscientious physician, who was urged in the medical literature to avoid hopeless cases, was encouraged to undertake difficult or uncertain ones. While opinions, both medical and lay, undoubtedly varied, the following statement by Celsus (first century A.D.) is quite typical of medical attitudes: "For it is the part of a prudent man first not to touch a case he cannot save, and not to risk the appearance of having killed one whose lot is but to die; next when there is grave fear without, however, absolute despair, to point out to the patient's relatives that hope is surrounded by difficulty, for then if the art is overcome by the malady, he may not seem to have been ignorant or mistaken" (*De Medicina* 5, 26, 1, C).

Although Christianity introduced a significant change in attitudes toward suffering in general and illness in particular, there appears to have been no correlative sense of duty to prolong life. The attitude toward death that is expressed in early Christian literature is that death is not to be sought, but life is not to be clung to. Augustine points to the irony that so many, when faced with troubles, cry out, " 'O God, send me death; hasten my days.' And when sickness comes they hasten to the physician, promising him money and rewards" ([17], p. 24). Augustine further laments "what things men do that they may live a few days. . . . If, on account of bodily disease, they should come into the hands of the physician and their health should be despaired of by all who examine them; if some physician capable of curing them should free them from this desperate state, how much do they promise? How much is given for an altogether uncertain result? To live a little while now, they will give up the sustenance of life" ([17], p. 25). But this is not what Christians are to do. They are to put their faith in God, leave the results in His hands and not cling to life with desperation. Basil of Caesarea expresses succinctly what seems to be a quite balanced position, accurately representative of the mainstream of Christian thought of his time. He writes: "Whatever requires an undue amount of thought or trouble or involves a large expenditure of effort and causes our whole life to revolve, as it were, around solicitude for the flesh must be avoided by Christians. Consequently, we must take great care to employ this medical art, if it should be necessary, not as making it wholly accountable for our state of health or illness, but as redounding to the glory of God. . . ." (*The Long Rule* 55). The idea of a duty to attempt to prolong life was not a correlate of the very strongly felt obligation to succor and comfort the sick and, generally, to preserve life until God should choose to end it. Indeed, the very principle of an obligation to prolong life, in any sense approaching modern conceptions, would likely have been viewed as being tantamount to blasphemy.

Whether to take on a hopeless case, or stay with a patient once his condition was assessed as hopeless, was sometimes a difficult and delicate choice for a physician. A good example of this is seen in the peculiar and testing circumstances of the various plagues that ravaged Europe for several centuries beginning in the mid-fourteenth century with the Black Death. Numerous treatises on the plagues were composed that have been recently analyzed with a view to gaining insights into the ethics of physicians in the midst of plague. We quote:

> Jacme d'Agramont (1348) stressed the highly contagious nature of the plague, mentioning cases where 'the master and the servants died of the same disease, and even the physician and the confessor.' He then wrote: 'Therefore all physicians should guard in times of pestilence, against financial cupidity, because he, who has such a motive, may bring about his own death and that of his friends. Unless he be the son of avarice and greed he would have given all the treasures of the world to avoid such a result.' The conscientious physician was in a delicate position in relation to public opinion that impugned his actions with charges of avarice if he seemed too eager to take on cases (especially if they terminated in death) and with charges of cowardice or irresponsibility if he were not willing to undertake the care of those ill with contagious disease. Some chroniclers living at the time of the Black Death complained that no amount of money could get physicians to treat the sick. Other physicians attempted to treat the sick without thought of remuneration. One physician, for example, wrote in his diary concerning a female patient who 'died of the worst and most contagious kind of plague, that of blood spitting,' that he treated her 'out of compassion as I would not have done it for money.' Quacks appeared to thrive during outbreaks of pestilence, moved by greed to promise recovery to the hopeless. . . . It was considered necessary to visit the patient to determine whether or not he was suffering from pestilence. If the condition were

diagnosed as plague, some physicians would then seek to determine whether or not the patient were curable. The author of an anonymous plague tractate composed in 1411 gives some interesting advice: '. . . if it is certain from the symptoms that it is actually pestilence that has afflicted the patient, the physician first must advise the patient to set himself right with God by making a will [and] by making a confession of his sins, as is set forth according to the Decretals. . . . Next the physician should examine the patient's urine and feces and take his pulse. If the patient is curable, the physician will undertake treatment in God's name. If he is incurable, the patient should leave him to die, in accord with the commentary on the second of the aphorisms. Those who are going to die must be distinguished by prognostic signs and then you should flee from them. He labors in vain who attempts to treat such as these.' The physician acted totally within the strictures of accepted ethics by refusing to treat a patient for whom the physician had no hope of recovery. The only criticism that contemporaries could make would be that he was in error to regard the condition as untreatable ([6], pp. 415–417).

But was the physician who, in the fifteenth century, refused to treat a patient for whom he had no hope of recovery, really acting 'totally within the strictures of accepted ethics'? This was a time in which attitudes toward physicians' responsibilities to the hopelessly ill were beginning to change. Tensions were strong; physicians who refused to treat patients were easily suspected of deserting them because the patients could not pay them enough. Similarly physicians who continued to treat such patients were just as easily accused of greed, in defrauding helpless patients. These two extremes are well illustrated by the following quotations from two sermons preached in England in the fourteenth century. The first is by Lanfranc: "O wretched physician, who for the money that you may not hope to get, desert the human body travailing in peril of death; and allow him, whom, according to the law of God, you should love and have most concern for, of all creatures under heaven, to be in jeopardy of life and limb, when you can and know how to apply a suitable remedy" ([29], p. 351). The second is by Bromyard: "All craftsmen would at once refuse a job for which unsuitable materials were provided. If a carpenter were offered wages for the building of a house with planks that were too short or otherwise unsuitable, he would at once say: 'I will not take the wage or have anything to do with it, because the timber is of no use.' Similarly the physician who can see no hope of saving his patient" ([29], p. 347).

The sentiments that Bromyard expressed were popular and deeply rooted. But attitudes were beginning to change, particularly among the moral theologians who, during the late Middle Ages and Renaissance, were probing casuistically into a wide range of ethical concerns. The fifteenth-century moral theologian Antoninus of Florence writes that "despite cases which, according to the judgments of men, are held to be fatal, sometimes the diligent physician is able to cure, but rarely. . . . Therefore, clear to the end the physician ought to do what he can to cure the patient." Should the physician receive a fee for such services if they are ineffective? Antoninus goes on to say that "because the physician was created as an instrument of nature, the instrument of medicine should not be entirely withdrawn from the patient as long as nature does not succumb. Therefore, the physician does not sin by accepting a stipend for the treatment of an illness which, following the principles of the art of medicine, he believes is incurable" ([3], p. 93).

By the late sixteenth century the simple desertion of the incurable by physicians must have become relatively rare. Francis Bacon writes that "in our times, the physicians make a kind of scruple and religion to stay with the patient after he is given up. . ." ([7], p. 28). He levels no criticism against their continued attendance on the terminally ill, but castigates physicians for failing to give more diligence (1) to finding or developing cures for conditions thought incurable, and (2) to alleviating the pain of the dying. This very basic change in attitudes toward the responsibility of a physician to his patient is neatly illustrated by contrasting positions of two popes, one from the late twelfth and the other from the mid-eighteenth century. Both have to do with the question of a physician entering holy orders. In the late twelfth century, Pope Clement III responded to a physician who requested admission to holy orders but was troubled that he may have unknowingly incurred, as a result of his medical practice, an irregularity that would, or should, if known, be an obstacle to his admission to orders. Clement replied that he should search his memory to ensure that he had never, even unintentionally, harmed a patient by any treatment that he had administered (*Decretales* 1, 14, 7). When Pope Benedict XIV addressed the same question in the mid-eighteenth century, he stated that physicians who wished to enter orders should first obtain a dispensation *ad cautelam*, i.e., as a precaution, since they never can be absolutely certain that they have always employed every means at their disposal for those patients who died under their care (*De Syn. Dioec.* 1, 13, 10). There is a vital change here: while in the twelfth century the concern was with possible harm inflicted on a patient actively, in the eighteenth it was with harm that resulted passively. In the first the question was, "Did you ever harm patients by the treatment you gave them?" In the second, "Did you ever harm patients by failing to give them the treatment you should have given?" These two documents by themselves prove nothing, but have been introduced here further to illustrate a very fundamental change both in physicians' sense of responsibility to their patients and in lay expectations, a change rooted, at least in part, in a redefining of medical practice from a right to a privilege.

THE PRACTICE OF MEDICINE: A RIGHT OR A PRIVILEGE?

We have already suggested that a major change in the patient-physician relationship occurs when magico-religious medicine ceases to be the primary means of health care and becomes simply an alternative to rational health care, and when physicians cease to be those endowed with a mysterious knowledge of magic and the supernatural and become those seen as skilled in a craft and, perhaps, knowledgeable of a science. A second, and nearly as significant, change occurs when the practice of medicine ceases to be a right and becomes a privilege. Throughout antiquity and the early Middle Ages, all who wished could call themselves physicians, engage in medical practice, and treat the ill or injured. There simply were no restrictions. In the late Middle Ages a change began to occur in various parts of Europe, a change brought about by movements both from outside and within the medical profession. It involved the imposition of licensure requirements by secular or ecclesiastical authorities in an attempt to protect the public from charlatans, and the organization of medical and surgical guilds (including universities) by practitioners in an effort to secure and protect a monopoly in providing medical and surgical services. The earliest datable imposition of medical licensure requirements by secular authority was by Roger II of Sicily in 1140, revised and made more stringent by his grandson, Frederick II, in 1231. In Montpellier papal legates in 1220 issued statutes for the university that included the stipulation that only those duly examined at the university were to be permitted to practice medicine in that area. These cases were somewhat exceptional, since throughout Europe the restricting of medical and surgical practice to those properly qualified was accomplished through guilds that sought to obtain and then protect a monopoly in their services. Whether the right to hold a monopoly was granted by ecclesiastical or secular authority, civic or regal, requests for a monopoly were always couched in terms describing the benefits that the community would derive if the guild be given the right to make and enforce standards of quality in its services, to control working conditions, to limit competition among members, to limit entry into the craft or profession, and to ensure the proper treatment of customers. One aspect of any monopoly is the right to train and, essentially, to license new members, thus eliminating competition from outside the guild. Although one of the major concerns in these measures was economic, yet the claim was frequently made by the guilds that such restrictions were necessary to maintain a high degree of competence and ethics in the craft or profession.

Medical or surgical guild regulations emphasize responsibilities to three different groups: (1) To the craft or profession. Practitioners are to do all things faithfully that pertain to their calling, to be willing and ready to help their colleagues with counsel or deed, and to be concerned for the well-being and honor of the guild. (2) To the people. They are faithfully to serve the people, to be available to attend those who need their services, and to charge them reasonable fees. (3) To the state. In a supervisory capacity, they are to report to the appropriate officials the failings of their colleagues. In a forensic capacity, they are to give truthful information to the authorities concerning the maimed, the wounded, and others. This tripartite responsibility – to the profession, to the people, and to the state – is the fundamental principle of medieval guild ethics.

The guilds were functional organizations that were designed to promote and protect the special interests of their members. They were brotherhoods, fraternities, companies of brethren united by a common economic activity, which was viewed as being best served by the subordination of individual interests to group or corporate interests. The goals, priorities, and obligations of the guilds were viewed as being of mutual benefit to the guild, the people, and the state. These benefits hinged upon the authority of the craftsmen or professionals to perform their functions unmolested by those who would illicitly meddle in their affairs. Thus an exclusive right to fill a particular role was sought and, in exchange for the privilege of holding a protected vocational status, a guild guaranteed as a *conditio sine qua non* a level of expertise in its services and the responsibility to police and to supervise its own members, both in respect to their qualifications and their performance. Regulations governing the minutiae of conduct, both within the guild and in relationships with customers (i.e., patients) and obligations to the state, varied considerably from guild to guild and from city to city. But the obligation to ensure competence and quality was a constant and essential feature.

When the practice of medicine is a right that all may exercise, responsibilities that impinge upon or temper the patient-physician relationship, except for physicians' liability for damage done to patients, are simply those that the individual physicians may consciously or unconsciously adopt and they are, ultimately, a product of their own moral principles. Since physicians are, of course, products of their environments, just as their patients are, the moral principles or ethics of the physician will, first and foremost, be the physician's ethics *qua* individual rather than *qua* physician, and only secondarily *qua* physician, and then tempered in great part by social expectations. This holds true whether medicine is practiced as a right or a privilege. However, individual conscience and social expectations are the major, perhaps the only, determinant when the practice of medicine is a right. These, of course, remain vitally important factors once the practice of medicine becomes a privilege. But social expectations and physicians' awareness of them are based upon a recognition that the privilege of practice is indeed a privilege, granted at the good pleasure of the society to be served by those to whom it is granted. This does not place the physician in the role of a public servant, dancing to the beck and call of the citizenry. Nor does it appear directly to affect many of the details of the relationship. The same tensions continue, the range of attitudes and prejudices remains the same, but the essential nature of the relationship changes in various ways, perhaps for the better in some ways (from the individual patient's perspective), perhaps for the worse. As long as the practice of medicine was a right, the patient-physician relationship was more personal in that it was not restricted or confined by professional or legal constraints. But patients were also much more at the mercy of physicians for whose competence no governing or regulating agency vouched, who were accountable to no peers for their conduct, procedures, fees, or anything else. It is no wonder that as long as the practice of medicine was a right, physicians' discretion either to accept a case or to refuse it on any grounds, or to leave a case if they chose, was criticized, if at all, only obliquely. But as the practice of medicine became increasingly a privilege, it was a privilege not simply granted to the individual but to all those collectively to whom the privilege had been granted individually: to corporate bodies of practitioners who assumed, corporately and individually, responsibilities for each other, to the patients (and all members of the community were potential patients), and to the authorities who granted and protected the privilege to practice. As a consequence, the options began to diminish, e.g., of refusing to treat a particular patient for a particular condition but of treating another patient for the same or a different condition, or of deserting a patient on the basis of the physician's personal judgment. Other options also were diminished, e.g., competition in fees. The privacy of the patient-physician relationship was lessened: obligations to report to the authorities various cases (e.g., of contagion, or of wounds caused by violence) became attached to the privilege of practicing the art.

MEDICAL ETIQUETTE

Only passing reference has been made above to medical etiquette. Medical etiquette is, of course, basically the deportment of physicians in their relations with their patients. The earliest writings on medical etiquette are from the pens of the authors of the earliest treatises on rational medical theory and practice, that is, the various authors of the Hippocratic Corpus. Partially in an attempt to establish and preserve a distinct professional identity tempered with dignity, and partially in recognition that certain types of conduct are inherently detrimental to practice, treatises, including the so-called Hippocratic Oath, were composed that set guidelines for conduct. Although the Oath differs from the rest of the Hippocratic Corpus in some aspects of ethics, e.g., abortion and euthanasia, in general principles of etiquette it is consistent with other deontological treatises. Most of the principles of etiquette are the product of common sense. Physicians should look healthy and not be of extreme weight. They should be 'perfect gentlemen', cheerful and serene in their dealings with patients, self-controlled, reserved, decisive, neither silly nor harsh. That physicians should not engage in sexual relations with patients or members of their households is stressed in several places. These basic principles of deportment, which are sometimes labelled 'Hippocratic ideals', were adopted by the authors of the deontological literature of the Middle Ages and are still regarded as central features of medical etiquette.

CONCLUSION

If we identify the expectations or hopes that modern patients usually have regarding physicians, there are four that are products of the period under consideration in this paper. First, chronologically, is the expectation that physicians are above all products of a scientific training and orientation, i.e., that they deal with disease and other physical ailments both empirically and rationally, not magically, mystically, or superstitiously. This expectation results from the Greek impact on medical theory and practice. Second, and chronologically concurrent with the first and also a product of Greek medicine, is the expectation that physicians are guided by certain basic standards of deportment or professional etiquette in dealing with patients. Third in time, and as a result of Christian influence, is the expectation, or at least hope, that physicians are compassionate and motivated, at least in part, by a genuine concern for their patients. Last is the expectation, generally taken for granted as much as the first, and this a product of the late Middle

Ages and Renaissance, that physicians are competent insofar as competence results from training and testing, and is vouched for by regulating authority.

Western Washington University, Bellingham, Washington, and
Oregon State University, Corvallis, Oregon

NOTES

[1] The use of non-gender designations, and the necessarily awkward circumlocutions that it requires, reflect the practice of the series and not that of the authors.

[2] Among many discussions of primitive medicine are [1], [2], [31], and ([33], Vol. 1).

[3] For medicine in ancient Egypt, see [18], ([33], Vol. 1), and [38].

[4] For medicine in ancient Mesopotamia, see [9], [30], and ([33], Vol. 1).

[5] For medicine in ancient Persia, see [10].

[6] See [4].

[7] There is an enormous literature on Greek and Roman medicine. On early Greek medicine ([33], Vol. 1) is very useful. The various writings of Edelstein [14], [15] and Temkin [34] are invaluable for both Greek (as well as Roman) medical theory and practice.

[8] For a discussion of the relationship of Christianity and secular medicine, see [12], [16], and [22].

[9] e.g., *The Rule of St. Benedict* 31; Cassiodorus, *Institutes of Divine and Human Readings* 1, 31.

[10] On medicine as a craft, see ([15], pp. 350–351 and 87–90; [34], pp. 137–153).

[11] See the example quoted from Demosthenes in ([19], pp. 165–166).

[12] See [24] and ([11], pp. 203–205).

[13] On philanthropy in the ancient world, see [21]. On philanthropic attitudes in ancient, early Christian, and medieval medicine, see [8].

[14] e.g., by Sir William Osler in *The Old Humanities and the New Science*, as quoted in ([15], pp. 319–320); and by ([25], pp. 21–22 and 245, Note 1).

[15] For a discussion of these motifs in classical literature, see [5].

BIBLIOGRAPHY

1. Ackerknecht, E. H.: 1942, 'Problems of Primitive Medicine', *Bulletin of the History of Medicine* **11**, 503–521.
2. Ackerknecht, E. H.: 1943, 'Psychopathology, Primitive Medicine, and Primitive Culture', *Bulletin of the History of Medicine* **14**, 30–67.
3. Amundsen, D. W.: 1981, 'Casuistry and Professional Obligations: The Regulation of Physicians by the Court of Conscience in the Late Middle Ages', *Transactions and Studies of the College of Physicians of Philadelphia* N. S. **3**, 29–39 and 93–112.
4. Amundsen, D. W.: 1978, 'History of Medical Ethics: Ancient Near East', in W. T. Reich (ed.), *The Encyclopedia of Bioethics*, Vol. 2, The Free Press, New York, pp. 880–884.

5. Amundsen, D. W.: 1977, 'Images of Physicians in Classical Times', *Journal of Popular Culture* **11**, 642–655.
6. Amundsen, D. W.: 1977, 'Medical Deontology and Pestilential Disease in the Late Middle Ages', *Journal of the History of Medicine and Allied Sciences* **32**, 403–421.
7. Amundsen, D. W.: 1978, 'The Physician's Obligation to Prolong Life: A Medical Duty without Classical Roots', *Hastings Center Report* 8 (August), 23–30.
8. Amundsen, D. W. and Ferngren, G. B.: 1982, 'Philanthropy in Medicine: Some Historical Perspectives', in E. E. Shelp (ed.), *Beneficence and Health Care*, D. Reidel Pub. Co., Dordrecht, Holland, pp. 1–31.
9. Biggs, R.: 1969, 'Medicine in Ancient Mesopotamia', *History of Science* **8**, 94–105.
10. Brandenburg, D.: 1972, 'Avesta und Medizin: Ein literaturgeschichtlicher Beitrag zur Heilkunde im alten Persien', *Janus* **59**, 269–307.
11. Crook, J.: 1967, *Law and Life of Rome*, Cornell University Press, Ithaca.
12. Dawe, V. G.: 1955, *The Attitude of the Ancient Church Toward Sickness and Health*, unpublished doctoral dissertation, Boston University School of Theology.
13. Durant, W.: 1953, *The Story of Civilization*, Vol. 5, Simon and Schuster, New York.
14. Edelstein, E. J. and Edelstein, L.: 1945, *Asclepius: A Collection and Interpretation of the Testimonies*, Vol. 2, The Johns Hopkins Press, Baltimore.
15. Edelstein, L.: 1967, *Ancient Medicine: Selected Papers of Ludwig Edelstein*, O. Temkin and C. L. Temkin (eds.), The Johns Hopkins Press, Baltimore.
16. Frost, E.: 1949, *Christian Healing*, Mowbray, London.
17. Getty, M. M.: 1931, *The Life of the North Africans as Revealed in the Sermons of Saint Augustine*, The Catholic University of America Press, Washington, D. C.
18. Ghalioungui, P.: 1963, *Magic and Medical Science in Ancient Egypt*, Hodder and Stoughton, London.
19. Glotz, G.: 1967, *Ancient Greece at Work*, translated by M. R. Dobie, W. W. Norton and Co., New York.
20. Hammond, E. A.: 1960, 'Incomes of Medieval English Doctors', *Journal of the History of Medicine and Allied Sciences* **15**, 154–169.
21. Hands, A. R.: 1968, *Charities and Social Aid in Greece and Rome*, Cornell University Press, Ithaca.
22. Kelsey, M. T.: 1973, *Healing and Christianity in Ancient Thought and Modern Times*, Harper and Row, New York.
23. Kibre, P.: 1953, 'The Faculty of Medicine at Paris, Charlatanism and Unlicensed Medical Practice in the Later Middle Ages', *Bulletin of the History of Medicine* **27**, 1–20.
24. Kudlien, F.: 1976, 'Medicine as a "Liberal Art" and the Question of the Physician's Income', *Journal of the History of Medicine and Allied Sciences* **31**, 448–459.
25. Lain Entralgo, P.: 1969, *Doctor and Patient*, translated by F. Partridge, McGraw-Hill, New York.
26. Marrou, H. I.: 1964, *A History of Education in Antiquity*, translated by G. Lamb, New American Library, New York.
27. Middleton, J. (ed.): 1967, *Magic, Witchcraft, and Curing*, Natural History Press, Garden City.
28. Mossé, C.: 1969, *The Ancient World at Work*, translated by Janet Lloyd, W. W. Norton and Co., New York.
29. Owst, G. R.: 1961, *Literature and Pulpit in Medieval England*, Blackwell, Oxford.

30. Ritter, E. K.: 1965, 'Magical Expert (=*Āšipu*) and Physician (=*Asû*): Notes on Two Complementary Professions in Babylonian Medicine', *Studies in Honor of Benno Landsberger on His Seventy-Fifth Birthday*, University of Chicago Press, Chicago, pp. 299–321.
31. Rivers, W. H. R.: 1924, *Medicine, Magic, and Religion*, Harcourt, Brace and Co., New York.
32. Sigerist, H.: 1943, *Civilization and Disease*, University of Chicago Press, Chicago.
33. Sigerist, H.: 1951, 1961, *A History of Medicine*, Vols. 1 and 2, Oxford University Press, New York.
34. Temkin, O.: 1977, *The Double Face of Janus and Other Essays in the History of Medicine*, The Johns Hopkins Press, Baltimore.
35. Thompson, S.: 1955–1958, *Motif-Index of Folk Literature*, University of Indiana Press, Bloomington.
36. Welborn, M. C.: 1938, 'The Long Tradition: A Study in Fourteenth-Century Medical Deontology', in J. Cate and E. Anderson (eds.), *Medieval and Historiographical Essays in Honor of James Westfall Thompson*, University of Chicago Press, Chicago, pp. 344–357.
37. Westermann, W. L.: 1955, *The Slave Systems of Greek and Roman Antiquity*, The American Philosophical Society, Philadelphia.
38. Wilson, J. A.: 1962, 'Medicine in Ancient Egypt', *Bulletin of the History of Medicine* **36**, 114–123.

LAURENCE B. McCULLOUGH

THE LEGACY OF MODERN ANGLO-AMERICAN MEDICAL ETHICS: CORRECTING SOME MISPERCEPTIONS

The purpose of this paper is to correct what I take to be three basic misperceptions about the history of Anglo-American medical ethics, the ethics of the patient-physician relationship, in the modern period (18th–19th centuries) and the legacy of that history to our century. The first two misperceptions are about the history of Anglo-American medical ethics in the two centuries preceding our own, First, historical medical ethics is taken by some to be little more than a collection of essays on medical etiquette and thus devoid of serious content *qua medical ethics* [6]. Ethics in medicine should, after all, have to do with something more than, say, the appropriate color and style of a physician's dress or fee-splitting in consultations. Second, medical ethical writings in the period under consideration are sometimes understood to be based in loosely held concepts about what a 'gentleman' was, a *Christian* gentleman in particular [7]. That is, historical medical ethics, including the work of British and American physicians of the period, has been understood to be devoid of serious philosophical content and to be devoid, as well, of roots in historical philosophical ethics.

The third misperception is more complex in character: it concerns the relationship between modern and contemporary Anglo-American medical ethics. It is reflected in what I call the 'standard account' of contemporary medical ethics. According to this account, ethical problems in medicine arise for patients and their families, for physicians, nurses, and other health professionals, for health care institutions, and for our society generally, because of the stupendous advances in medical technologies. We now have ultrasound scanning and computerized tomography, exquisite techniques in cardiac and brain surgery, respirators and 'crash-carts' – and they threaten to overwhelm us. In short, we live in an era of rapid and profound change in medicine – as in so many other aspects of our lives – change that threatens to outstrip our moral sensibility. Because of the misperceptions outlined above, contemporary medical ethics has moved forward on the assumption that historical medical ethics provides us no useful resources. Hence, we have nothing to learn from that history – except not to repeat its mistakes – and so we are free to inquire into the ethical dimensions of the patient-physician relationship as if that history didn't matter. This is the third misperception which this paper will attempt to correct.

Earl E. Shelp (ed.), The Clinical Encounter, 47–63.

With respect to the first misperception, I shall argue that medical ethics in the period under consideration included more than simply etiquette. There can be no doubt that writing about medical etiquette consumed the energies of various writers and medical associations. Even a cursory glance at medical ethical codes – which sometimes address even such matters as fisticuffs at meetings and dueling – sustains this view. The roots of this misperception lie deeper, however. They are to be found in what has come to be the canon of modern medical ethics: Percival's *Medical Ethics* [25] in 1803, followed by the first *Code of Ethics* of the American Medical Association in 1847 [1]. Those two documents are surely central to the history of modern Anglo-American medical ethics but the *overemphasis* on them has obscured our vision of the past. Physicians were concerned with a great deal more than simply medical etiquette, as writers of the period themselves observed [12, 26]. Instead physicians gave serious attention to such matters as truthtelling to the dying [14] and experimentation with human subjects [13].

As physicians turned to such medical ethical topics they drew on a wide variety of sources. No doubt as Chester Burns has shown [7], religion was among those sources. It is a mistake, however, to take religious insight to be a sole – or even major – foundation for modern medical ethics. Philosophical ethics was also influential, especially that of the Scottish moralists. Practitioners of contemporary medical ethics have failed to appreciate this feature of modern medical ethics – which is not surprising given the Percival-AMA Code canon. Thus, in order to correct the second misperception, we need to expand the canon of modern Anglo-American medical ethics.

Once the expansion has been undertaken, the third misperception is called into question. Serious, philosophically oriented efforts in medical ethics may serve as valuable resources, even precedents for contemporary efforts. I have addressed this feature of modern medical ethics elsewhere, and so it will not be my main concern here [21]. Instead, to use the metaphor of Alasdair MacIntyre [20], the legacy of modern medical ethics to our century is a series of fragments of what I call the 'beneficence model' of the ethical dimensions of the patient-physician relationship [23]. This model was given philosophical formulations in the modern period, but those formulations dissolved and only pieces of them come down to us. This fragmentation of medical ethics, I shall argue, is of perhaps greater import than the rise of medical technologies for how the central task of contemporary medical ethics, understanding the ethical dimensions of the patient-physician relationship, should be understood.

MEDICAL ETHICS AND MEDICAL ETIQUETTE

If one read only Percival's *Medical Ethics*, it would not be difficult to conclude that, from the eighteenth to the nineteenth centuries, modern medical ethics was equivalent with, by being reduced to, medical etiquette. His text opens with a consideration of 'Professional Conduct, Relative to Hospitals, or Other Medical Charities' [25]. There is some mention of the need for secrecy and confidentiality in the patient-physician relationship, as well as not misleading the dying as to their prognoses. But Percival quickly moves on to such matters as consultations, the distinction between medical and surgical cases, the content of hospital registers, and the like including – later in the text – dueling. Thus medical *ethics* is given short shrift. The 1847 AMA *Code of Ethics* – the other half of the standard canon – includes sections on the obligations of physicians to their patients. These, however, merely rehearse Percival's remarks. The bulk of the *Code* is given over to the intra-professional concerns, such as treating colleagues' family members, consultation, and conduct in 'cases of interference', for example, sudden emergencies requiring attention before the patient's regular physician arrives [1].

There is no good reason, however, to limit the canon to these two texts. To be sure, Percival's little book was enormously influential, especially on the AMA *Code*. And the *Code* was surely influential because it was the centerpiece of mid-nineteenth century efforts to organize the medical profession on a national basis in the United States. But there are other widely influential writings, in particular the *Lectures on the Duties and Qualifications of a Physician* by Dr. John Gregory of Edinburgh [14]. Gregory's concern was not simply with matters of medical etiquette. Indeed his inquiry ranged quite broadly, considering such topics as the patient's role in choice of treatment, truth-telling to the sick and dying, and the obligations not to abandon hopeless cases. And, when he does address matters of etiquette – consultations, relations with surgeons and apothecaries, dress, the appropriate number of patient visits, and "little peculiarities" of manners ([14], p. 62) – he does so in terms of the same ethical principles that he employed in discussing medical ethical issues. In America, we find other examples of medical ethics as we know it, namely the work of Samuel Bard [3] in the eighteenth century and Austin Flint in the nineteenth [12]. John Gregory's medical ethics have been examined in detail elsewhere [21] and so Samuel Bard's will be considered in detail below. For now, suffice it to say that the first misperception – that medical ethics previous to our own century was preoccupied with medical etiquette – can be sustained only if we artificially constrain the canon of historical medical ethics. But there is no good reason – historical or contemporary – for doing so. Hence, this first misperception can be quite easily corrected.

MEDICAL ETHICS AND PHILOSOPHICAL ETHICS

The second misperception is the view that modern medical ethics lacks a serious philosophical character, because it is based on professional concerns [27] or draws on religious sources [7]. To be sure, there are guild elements in modern medical ethics and the influence of Christianity is evident in many writings. Hence, like the first, this second misperception is partly true but it is incomplete in that it blinds us to the influence of modern philosophical ethics on modern medical ethics.

While the complete history of modern medical ethics remains to be written, it is possible to cite examples of the influence of moral philosophy on medical ethics. One such is the work of John Gregory [14, 21]. Another is the work of Samuel Bard, a major figure in American medicine in the last third of the eighteenth century. I want now to consider one of his essays, *A Discourse Upon the Duties of a Physician* . . . [3]. This essay was delivered as an address at the first medical commencement held at King's College in New York City in 1769. Like many physicians in America in the eighteenth century Samuel Bard was educated abroad, at the medical school in the University of Edinburgh, one of the great medical schools of its day. Bard returned to his native soil to help found the second American medical school, King's College, wherein he was "appointed to teach the theory and practice of Physic, the most important branch of all" ([9], p. 509).

At first glance Bard's commencement address seems to be little more than a treatise on medical etiquette. On a second – and more sustained – look, however, it shows itself to be more than that. In a series of greatly condensed paragraphs – the occasion, obviously, not being one for a lengthy didactic treatise – Bard addresses a number of ethical issues in the patient-physician relationship, e.g., the physician's obligations to the dying patient, the need to remain educated and up-to-date, the proper disposition to maintain toward those who are ill, and the obligations of individual physicians and the medical profession at large to the sick poor. What this closer look reveals is a feature of Bard's essay that, to the best of my knowledge, has escaped the notice of medical historians and historians of medical ethics: the influence of Scottish moral philosophy on the formation of Bard's views. Located more exactly within the tradition of the Scottish moralists, Bard's indebtedness seems to be to Francis Hutcheson [15]. That is, Bard makes his arguments, such as they are, about the duties physicians should have to their patients in terms of the Hutchesonian concepts of Duty and Benevolence and does so in an explicit way, even using the language of Hutcheson's moral philosophy. Moreever, Bard treats matters of medical etiquette in this same mode, providing for them a serious philosophical foundation and context.

The Bearing of Hutcheson's Moral Philosophy

It is difficult to be sure exactly why it is that this character of Bard's medical ethics has been overlooked. Two possible explanations come to mind. The first may be that the manner in which we educate physicians today is thought to be a model of long-standing. That is, we do not today assume that our physicians are broadly trained in the humanities, though, happily, this trend is being altered to some extent. Students of medicine, even so, are occupied for the most part with the biomedical sciences and with the clinical practice of scientific medicine. In the eighteenth century this was not the case, if only because there was as yet so little basic and clinical science to know. As opposed to surgeons, scurrilously referred to in the same breath with barbers, who were trained for the most part by apprenticeship, physicians were, when university educated, educated in more than simply science. Boorstin summarizes nicely the sort of education a physician would have received at a school like Edinburgh.

> Necessarily a master of the classical languages in which medical knowledge for the past had been presented, he was also a man of general learning. Thus, when Henry VIII chartered the Royal College of Physicians in 1518, he intended to set up both a learned academy and an exclusive guild for these practitioners of 'physick' ([5], p.228).

The second possible explanation is connected to the first. Part of the education of the members of a 'learned academy' would surely include, if not be centered in, philosophy. And, the great philosophers of the day in Great Britain included Francis Hutcheson, whose influence was surely widespread. Given such an academic milieu it should be expected that the thinking of university students in Great Britain would be influenced accordingly. Yet, such a possible source of influence on the formation of the medical ethics of physicians like Samuel Bard has been ignored.

That influence was probably also felt closer to home. In his probing and provocative study, *Inventing America*, Professor Gary Wills is at great pains to counterbalance the view that it was John Locke's moral and political philosophy that alone influenced Thomas Jefferson in the preparation of his drafts of the *Declaration of Independence* [28]. Instead, it was Scottish moralism and Hutcheson's philosophy in particular that, according to Wills, should be recognized as profoundly influential on Jefferson's thought. I am not really in a position to judge the matter of Hutcheson's influence on Jefferson and, in any case, this particular controversy is not of direct concern here. The more general remarks by Wills about Hutcheson's influence in American moral thought are, however, pertinent. As Professor Wills puts it:

What is often said about Locke's politics was literally true of Hutcheson's ethics – it was 'in the air,' making up an intellectual 'atmosphere,' a kind of tacit orthodoxy. Hutcheson was *the* author of this central topic of philosophy [i.e., moral philosophy] as it was taught at Philadelphia and Princeton and New York ([28], p. 201).

If what follows is correct, then Wills's view on the pervasiveness of this atmosphere is given further corroboration.

Hutcheson on the Moral Sense, Benevolence, and Duty

At the core of Hutcheson's moral philosophy, a philosophy Bard adapted to his own circumstances, is the moral sense, a concept developed at length in Hutcheson's *An Inquiry into the Original of our Ideas of Beauty and Virtue* [15]. In a marvelous piece of metaphysical anthropology – he means to make a claim about a fundamental character of human nature – Hutcheson describes the true spring "of those actions which are counted virtues" ([15], p. 143). This wellspring of virtue is "some determination of our nature to study the good of others; or some instinct, antecedent to all Reason for interest, which influences us to the love of others" ([15], p. 143). To this source of virtue is coupled the moral sense, another fundamental feature of human nature, which "determines us to approve the actions which flow from this love in ourselves to others" ([15], p. 143). Thus, we do not act virtuously when we act selfishly, in an exclusively self-regarding manner. Not only is this the case, but it is also a fact of human nature that we are inclined toward other-regarding actions: actions whose ends or purposes are not construed in terms of our own interests but, exclusively it would seem, in terms of the interests, needs, wants or whatever of others. Moreover, the moral sense – a feature of 'universal extant' – moves us to approve other-regarding actions. In other words, to approve an exclusively self-regarding action as virtuous is contradictory, revealing that one does not – to use the terminology of Wittgenstein [29] – know the language game of the word 'virtuous'. Thus, the pinnacle of right action is virtuous action, action that Hutcheson says is motivated by benevolence. Indeed, the perfection of virtue or benevolence is when "a being acts to the utmost of its power for the *publick good*" ([15], p.172).

The last feature of Hutcheson's moral theory that is relevant here is his treatment of obligation. In its proper sense the term, 'obligation', means

> . . . a determination, without regard to our own interest, to approve actions, and to perform them; which determination shall also make us displeased with ourselves, and uneasy upon having acted contrary to this sense of obligation ([15], p. 249).

Not surprisingly, the paradigm case of moral obligation is the act founded in benevolence: ". . . in this meaning of the word obligation, there is naturally an obligation upon all men to benevolence" ([15], p. 249).

Hutcheson elaborates on these basic concepts of his moral philosophy in the *Inquiry* and his final position amounts to the following. A particular sort of pleasure or pain accompanies our actions. Pleasure, by definition it would seem, accompanies those actions that are other-regarding, while pain accompanies self-regarding actions. These, it would seem, are inescapable features of what it is to be a human being and thus a moral agent. To act benevolently, then, is to act in such a way that one enjoys the pleasure that accompanies self-less, other-regarding actions. To act malevolently is to act in such a way that one suffers the pain that accompanies selfish, self-regarding actions. The connection to obligation is made thusly: if an action is accompanied by pleasure of the required sort, we judge that act to be good or benevolent and, therefore, obligatory. If, on the other hand, an action is accompanied by pain, we judge it to be bad, hence prohibited. One ought to promote the good or interests of others, the "publick good," and not our own goods and interests. Put simply, one ought to do what is benevolent and ought not to do what is evil, the neglect of the public good.[1]

Bard's Employment of Hutcheson's Moral Philosophy

Now, Bard applies Hutcheson's moral philosophy in a fairly thoroughgoing and, I think, self-conscious manner to medical practice. The most striking examples of this approach to medical ethics are his remarks about the care of the dying patient and of the sick poor. The following analysis focuses on these sections of his essay, with a view toward dispelling the misperceptions noted at the beginning of this essay.

Bard's basic strategy is to have his students develop a "particular sensibility of disposition . . . which I believe is the best counterpoise to self-interest" ([3], p. 9). That is, one is to act always out of other-regarding benevolence and not out of selfish motivations. In treating the dying patient, for example, this counterpoise is necessary because self-interest (say, avoiding one's own fear of death or the feeling of having failed) may lead one to give false hope to the dying patient as a consequence of not being truthful with the patient. Even though such withholding of information is "at best a good natured and humane deception" ([3], p. 10), such a course of action has a darker side, one founded in the worse aspects of the self-interest of the physician, "the baser motives of lucre and avarice" ([3], p. 11), motives to be satisfied by further charging the patient for visits when nothing medically can be done to prevent impending death. Clearly, Bard is no mincer of words.

Bard's argument here is based squarely in a consideration of other-regarding actions and the satisfaction that devolves upon those who perform such actions. One should tell the patient that he or she is dying because not doing so is cruel, because the patient will be forewarned and thus able to prepare him or herself for death, and because the patient will be able to prepare his or her family for the death that is soon to come. Bard's language on this subject is quite moving. That is, it is meant to illustrate benevolence at work.

> . . . besides it [not telling the patient about impending death] is really cruel, as the stroke of death is always most keenly felt, when unexpected; and the grim tyrant may in general be disarmed of his terrors, and rendered familiar to the most timid, and apprehensive; either by frequent meditation, by the arguments of philosophy, or by the hopes and promises of religion. But even overlooking the important concerns of futurity; the business of this life may render such conduct [*i.e.*, not telling dying patients of their death] highly dangerous and criminal; as those to whom the thoughts of death are painful, are too apt when flattered with the prospects of recovery, to neglect the necessary provisions against a disappointment, and by this means involve their families in confusion and distress ([3], p. 11).

Clearly the pleasure that accompanies benevolent actions will be experienced by a physician who is honest with his or her dying patients.

What is going on here, I think, is this. Bard is trying to induce in his audience an experience of the pleasure that accompanies benevolence, so that they will recognize it when they experience it. The features of that experience, though, are not fabricated by Bard. Instead, his portrayal of them is deeply rooted in his own clinical experience of what has happened when doctors are less than fully honest with their dying patients. This is, then, no mere theoretical and simply rhetorical exercise; it is medical ethics solidly grounded in both philosophical ethics and clinical experience. No appeal is made to concepts of 'gentleman' or religious values or matters of medical etiquette, *simpliciter*. This is Enlightenment thought at its best.

An even more vivid demonstration of this facet of Bard's medical ethics occurs in the course of his argument that physicians have special obligations to the sick poor.

Let those who are at once the unhappy victims, both of poverty and disease, claim your particular attention; I cannot represent to myself a more real object of charity, than a poor man with perhaps a helpless family, labouring under the complicated miseries of sickness and penury. Paint to yourselves the agonizing feelings of a parent whilst labouring under some painful disease, he beholds a helpless offspring around his bed, in want of the necessities of nature; imagine the despair of an affectionate wife and a tender mother, who can neither relieve the pain and anxiety of her husband, nor supply the important cravings of her children; and *then* deny them your assistance if you can ([3], pp. 14–15).

Only a true moral wretch, someone wholly bound up in self-interest, could comfortably do so. But to be this sort of person cuts deeply against the grain of humanity, at least as Bard and Hutcheson would have us understand that state. Thus, Bard concludes:

... but the supposition is injurious to humanity, and *you* in particular, I know want no such incitements to duty and benevolence ([3], p. 15).

The indebtedness to Hutcheson is now plain. Benevolence, the perfection of virtue, requires that one serve the "publick good." For the physicians in New York City at the end of the eighteenth century, according to Bard, this means an obligation to treat the sick poor, regardless of their means to remunerate one for one's service.

One's benevolent concern for the public good should not, however, stop at the individual level; physicians should also support and contribute to the establishment of a public hospital in which to tend to the health care needs of the sick poor. After all, Bard argues, the labors of the poor earn them only enough to provide for the necessities of daily life, so sickness for them is indeed a 'calamity' ([3], p. 16); it causes them and their families to lose the very means of survival. Again, only the most morally callous would deny the rightness of doing all one can to ameliorate such calamities. Indeed, following the logic of Hutcheson's moral philosophy, there is an *obligation* to benevolence here, at both the individual and social levels. Thus, benevolence on the part of physicians for the sick poor reaches beyond the confines of the patient-physician relationship – tending to the sick poor without fee – to the broader public good – opening a public hospital and contributing to it both money and services.

The founding of such a hospital would not be without its practical consequences and, hence, further justification. Bard notes two sorts of consequences. First, by gathering the sick of the city together into one place, one would be in an excellent position to study diseases and their natural histories, and thus better understand the effectiveness of different antidotes to disease ([3], pp. 16–17). Second – and Bard says it is by no means the least of his arguments – such a hospital would provide "the best and only means of properly instructing pupils in the practice of medicine" ([3], p. 17). To this feature Bard adds in the second instance a final argument from benevolence. A public hospital would permit and sustain "the breeding of good and able physicians" ([3], p. 17), an enterprise in which society has, obviously, a high stake. Not only would such a hospital serve the public good by meeting the health care needs of the sick poor, it would also do so by providing for the disciplined training of future physicians – hence the balance of the title of Bard's address: ... *with some sentiments on the usefulness and necessity of a public hospital* [3].

Thus, as a corrective to the second misperception, I hope to have shown that modern medical ethics, on occasion, achieved considerable philosophical sophistication in its treatment of the complex dimensions of the patient-physician relationship. This achievement of eighteenth century physicians like Bard and Gregory, the placing of medical ethics on a philosophical foundation, however, was replaced by another influence in the nineteenth century: the trend toward etiquette and codification. This 'inward turning' of medical ethics, to focus on intraprofessional concerns, in the nineteenth century contributed to the fragmentation of medical ethics, a theme to which we now turn.

THE FRAGMENTATION OF MODERN MEDICAL ETHICS

If the preceding historical analysis is correct, then one of the major achievements of the modern period of Anglo-American medical ethics is the application of moral philosophy to the ethical dimensions of the patient-physician relationship. The result was a philosophical account of what I call the 'beneficence model' of the relationship [23]. This model focuses on the physician's moral character and the commitments proper to it. Its earliest expression is the Hippocratic Oath. This oath, like all such verbal acts, is a solemn promise about how one will employ medical knowledge and skills. Swearing an oath is thus more than a mere declarative utterance. It is also more than a performative utterance. It is a verbal act through which one transforms oneself.

This transformation comprises the three distinct moments of the professional model of the patient-physician relationship. The first is one's individual moral character, constituted by an individual's moral history and convictions. The second moment is an account of the proper end of medicine, the moral purposes one should serve as a physician. The final moment is a commitment to these moral purposes, a commitment that transforms subjective beliefs and convictions into the *professional* character of the physician. One's central moral convictions as a physician are thus shaped in terms of the proper end of medicine.

Alasdair MacIntyre provides a more general account of this transforming character of moral commitment [20]. As MacIntyre sees it, ethics properly understood consists of three elements. The first is "human nature in its untutored state" ([20], pp. 50–51): human nature as we happen to find it full of potentialities not yet ordered coherently toward proper human goals. The second element is "a conception of the precepts of rational ethics" ([20], p. 51): ethical principles organized into a coherent theory that show how properly to achieve human goals. The final element is the transformation of the first that is achieved by submitting to the discipline of the second: "the conception of human-nature-as-it-would-be-if-it-realized-its-*telos*" ([20], p. 51), its proper end or goal. MacIntyre finds this model of ethics operative in philosophers as diverse as Aristotle, Maimonides, Ibn Roschd, and Thomas Aquinas ([20], p. 51).

There is a direct parallel between the conceptual structure of the beneficence model of medical ethics and MacIntyre's account of ethics generally. Instead of a general account of the human *telos*, the beneficence model of medical ethics focuses on the *telos* of medicine. It is the commitment to this *telos* that makes one a physician *qua* professional. For Bard, as we have seen, the major end of medicine is patient-centered and shaped by benevolence and duty. All other concerns are secondary, including intraprofessional matters:

Never affect to despise a man for want of a regular education, and treat even harmless ignorance, with delicacy and compassion, but when you meet with it joined with foolhardiness and presumption, you may give it no quarter ([3], p. 9).

and medical etiquette:

Do not pretend to secrets, panaceas, and nostrums, they are illiberal, dishonest, and inconsistent with your characters, as gentlemen and physicians, and your duty as men – For if you are possessed with any valuable remedy, it is undoubtedly your duty to divulge it, that as many as possible may reap the benefit of it; and if not [which is generally the case] you are propagating a falsehood, and imposing on mankind ([3], p. 10).

In short, Bard's rendition of the professional model asserts that the proper goal of medicine is to benefit patients, while avoiding harm to them – echoing the classical formulation of the Hippocratic Oath and accompanying texts such as *Epidemics* [16]. John Gregory's *Medical Ethics* is similar: it emphasizes the obligation of the physician to "relieve the distresses" of their fellow creatures, namely the pain and suffering of disease ([14], pp. 20–21).

In the nineteenth century medical ethics seems to have abandoned this patient-centered account of the ethics of the patient-physician relationship, one that looked outward, as it were, and removes intraprofessional concerns to a secondary status. The reasons for this change are complex and a full treatment of them is beyond the scope of this paper. Jeffrey Berlant, for example, has identified the drive toward monopolization of medical care as a major reason why an approach like Percival's took such strong hold [4]. Donald Konold has identified a related reason, an interest in enhancing the prestige of the profession [19], a concern also found in the Hippocratic texts [10].

In short, medical ethics in the century preceding our own experienced a 'turning inward' in which intraprofessional concerns supplanted the patient-centered approach of the philosophically-oriented efforts in the preceding century. The result of this radical shift in focus was to unhinge medical ethics from the fledgling efforts to ground it in moral philosophy. That is, this turning inward deprived medical ethics of an intellectual rationale, by substituting professional self-interest for philosophical reasoning. It was not long before opposition to the codification of medical ethics began [26], resulting in a series of revisions of the AMA *Code* [2]. Indeed, a detailed code was eliminated in favor of statements of general principles, around which consensus could be formed [2]. By this time, the early decades of our own century, the philosophical foundations of medical ethics had become obscured. There is, for example, reference to the 'honored ideals' of the medical profession in various formulations of the AMA's principles, but the latest version of those principles abandons even this vague reference [2].

The result of abandoning philosophical foundations for medical ethics is that we can no longer say univocally what the proper moral end of medicine is. Recall that an understanding of that end is at the core of the professional model, as expressed by physicians like Bard and Gregory. Is medicine to pursue health? If so, is it to adopt the understanding of health as espoused by the World Health Organization or, by contrast, the definitions offered by Callahan [8] and Kass [18]? Is medicine to be without limits and wage war on disease and death, or is it to be more modest and limited in its character

[11]? Is the physician first to 'do no harm' or first to benefit patients [24]? By turning its vision inward, medical ethics in the nineteenth century lost the ability to answer such questions with authority, because it lost sight of the moral principles that should shape the proper end of medicine and thus the ethics of the patient-physician relationship.

Now, one might hope to escape this problem by reviving the eighteenth century project in medical ethics. On closer analysis, however, even a philosophically-grounded version of the beneficence model is subject to a similar fate, though for different reasons. If one were to follow Bard and Gregory and hold that the proper end of medicine is to benefit patients, while avoiding unnecessary harm to them, we would have made some progress over the nineteenth century, to be sure. But we would still have to say what will count as a benefit or a harm. Relieving the pain and suffering of disease is a benefit. Avoiding premature death and disability is another. Prolonging life is yet another, while prolonging life at any cost completes this rough continuum of benefits for patients. Where along this continuum could we finally locate the *telos* of medicine – as MacIntyre would have us do – if we are to avoid fragmentation?

In my view there is no finally satisfactory answer to this question. The reason for this is not simply the narrowing of medical ethical vision in the nineteenth century. Instead, the reason is found in the beneficence model of medical ethics itself. That model is correct to insist on the transforming character of the professional commitment of the physician to the welfare of his or her patient and society generally. But that model cannot provide a single account of the good of patients and thus of medicine, if only because inevitably the values of patients must be taken into account in determining what is good for them. Once this feature of medical ethics is recognized, then competing accounts of the *telos* of medicine are inevitable. Thus, even if the philosophically grounded medical ethics of the eighteenth century had come to dominate the nineteenth century, the legacy of modern medical ethics to our century would have been the same: the fragmentation of medical ethics.

Here, MacIntyre's analysis of contemporary philosophical ethics applies, *mutatis mutandis*, to contemporary medical ethics. According to MacIntyre, the second of the three elements of an adequate ethical theory, identified above, is missing in contemporary ethics. We have generally neglected the topic of proper human goals or where ethics has not neglected this topic, the result has been a series of competing and flawed accounts. Ethics has fragmented: we can no longer give an account of those moral principles that should shape human nature. Hence, the legacy of modern ethics is fragmentation: on the one hand, arguments about subjective moral convictions and values and, on the other, arguments about objective moral principles that should shape those convictions, with no account of proper human goals to join the two into an adequate ethical theory ([20], pp. 49–75).

If my argument so far is correct, there is an exact parallel between contemporary ethics and contemporary medical ethics, as expressed in the beneficence model of the patient-physician relationship. Because of the many competing accounts of the proper goal or *telos* of medicine, we have no longer the ability to join pre-professional, subjective values and convictions to professional and objective values and convictions in a way that would result in an adequate account of the obligations of the physician *qua* physician – precisely the goal of the beneficence model of the patient-physician relationship. Thus, that model has fragmented and it is those unjoined fragments that we inherit from the modern period in medical ethics.

Why is this an important consideration? The answer lies in a recognition that there are no issues in contemporary medical ethics that can be separated from debate about the *telos* of medicine. Issues as diverse as confidentiality in psychotherapy and discontinuing treatment for non-competent patients (infants, adolescents, and adults alike) all involve this debate. Indeed, alternative resolutions of debates on particular issues entail diverse views on what the *telos* of medicine should be. Holding that confidentiality is an absolute obligation in psychotherapy entails that the *telos* of medicine involves only the good of the individual patient. Holding that confidentiality is a *prima facie* obligation entails that the *telos* of medicine involves the good of the individual patient and the good of others whose welfare (lives and property) might be adversely affected by maintaining confidentiality. Rejecting quality of life judgments in cases of discontinuing treatment for non-competent, terminally ill patients entails the view that life itself is the fundamental good that medicine should seek. Admitting quality of life judgments as relevant in such cases entails the view that life under some conditions is the fundamental good that medicine should seek. In short, there is a systematic character to medical ethics, one that joins specific topics to general ethical themes and to each other through accounts of the proper goal of medicine.

That we cannot univocally conceive *the telos* of medicine – that we now must speak of multiple *teloi* of medicine – is the fundamental problem in contemporary medical ethics. As a consequence, the impact of rapid change and technological development must be understood differently. It is not that such change in technology outstrips our moral sensibility. Indeed, it does not, since we *can* give accounts, though only fragmentary ones, of the moral dimensions of that technology. It is that, concerning the *telos* of medicine, there is no single sensibility to be outstripped – there are many such sensibilities. Technological and other changes, therefore, only point to the deeper problem, that of pluralism for how we shall understand in systematic ways the ethical dimensions of the patient-physician relationship – the first part of the legacy of modern Anglo-American medical ethics to our century.

CONCLUSION

In this paper I set out to show why three views concerning the history of modern Anglo-American medical ethics are, in fact, misperceptions. If my analysis of them is correct, these misperceptions are not thoroughgoing. Each is partly true, but leaves out a crucial feature of that history. The overall aim of my investigation has been to show how contemporary medical ethics is distinctively and inescapably shaped by its history.

Three lessons can be drawn from the preceding study. First, by recognizing that in the past medical etiquette was sometimes taken seriously as an *ethical* issue, we come to see that contemporary medical ethics needs to expand its agenda. By doing so, it will come to appreciate how intraprofessional concerns and matters of etiquette are not only intrinsically interesting, they frequently are the cause of ethical dilemmas in medicine. Hence, they require closer attention. Second, studying the second misperception reminds us that hubris is an occupational hazard in contemporary medical ethics. We are decidedly not the first to attempt the wedding of philosophy and medicine. Past efforts remind us forcefully that this is so and they may well serve as guides for contemporary projects. Finally, analysis of the third misperception requires us to reconceive the basic task of contemporary medical ethics. It is not to study the ethical dimensions of technological change and its impact on the patient-physician relationship. Instead, we need to look deeper, at our moral pluralism and how that feature of our culture shapes the culture of the patient-physician relationship. By doing so, we shall appreciate even more how this or that resolution of a particular issue entails a particular view of a *telos* of medicine. In this way we will come to see that issues do not occur in isolation but in fact are joined together through our attempts to resolve them in terms of an account of that *telos* of medicine. In this way we will come to see that issues do not occur in isolation but in fact are joined together through our attempts to resolve them in terms of an account of that *telos*. This recognition, in turn, will require us to abandon a piecemeal approach to medical ethics in favor of a more systematic one – precisely the strategy of our forebears in the eighteenth century, the second part of the legacy of modern medical ethics to our century.

Georgetown University School of Medicine,
Washington, D. C.

NOTES

[1] Hutcheson's formulation is the following: "The degree of moral evil, or the degree of vice, which is equal to the hatred or neglect of the publick good. . ." ([16], p. 173).

BIBLIOGRAPHY

1. American Medical Association: 1847, 'Code of Medical Ethics', *Proceedings of the National Medical Convention* 1846–1847, pp. 83–106.
2. American Medical Association: 1981, *Current Opinions of the Judicial Council of the American Medical Association*, Chicago.
3. Bard, S.: 1769, *A Discourse Upon the Duties of a Physician With Some Sentiments on the Usefulness and Necessity of a Publick Hospital*, A. and J. Robertson, New York.
4. Berlant, J.: 1975, *Profession and Monopoly: A Study of Medicine in the United States and Great Britain*, University of California Press, Berkeley.
5. Boorstin, D.: 1958, *The Americans: The Colonial Experience*, Vintage Press, New York.
6. Branson, R.: 1976, 'The Scope of Bioethics: Individual and Social', in R. M. Veatch and R. Branson (eds.), *Ethics and Health Policy*, Ballinger, Cambridge, Massachusetts, pp. 5–16.
7. Burns, C.: 1978, 'North America: Seventeenth to Nineteenth Century', s. v. 'Medical Ethics: History of,' in W. T. Reich (ed.), *Encyclopedia of Bioethics*, Vol. 3, Macmillian, Free Press, New York, pp. 963–968.
8. Callahan, D.: 1976, 'Biomedical Progress and the Limits of Human Health', in R. M. Veatch and R. Branson (eds.), *Ethics and Health Policy*, Ballinger, Cambridge, Massachusetts, pp. 157–165.
9. Ducachet, H.: 1821, 'A Biographical Memoir of Samuel Bard', *The American Medical Recorder* 4, 609–633.
10. Edelstein, L.: 1967, *Ancient Medicine: Selected Papers of Ludwig Edelstein*, Johns Hopkins Press, Baltimore.
11. Engelhardt, H. T., Jr.: 1975, 'The Counsels of Finitude', in P. Steinfels and R. M. Veatch (eds.), *Death Inside Out*, Harper and Row, New York, pp. 115–128.
12. Flint, A.: 1883, *Medical Ethics and Etiquette*, Appleton and Company, New York.
13. Gregory, James: 1800, *Memorial to the Managers of the Edinburgh Infirmary*, Murray and Cochrane, Edinburgh.

14. Gregory, John: 1772, *Lectures on the Duties and Qualifications of a Physician*, W. Strahan, London.
15. Hutcheson, F.: 1971, *An Inquiry Into the Original of Our Ideas of Beauty and Virtue*, G. Olms Verlag, Hildesheim (original published in 1725).
16. Jones, W. H. S. (trans.): 1923, *Hippocrates*, Loeb Classical Library, Harvard University Press, Cambridge.
17. Jonsen, A. and Hellegers, A.: 1976, 'Conceptual Foundation for an Ethic of Health Care', in R. M. Veatch and R. Branson (eds.), *Ethics and Health Policy*, Ballinger, Cambridge, Massachusetts, pp. 17–23.
18. Kass, L.: 1975, 'Regarding the End of Medicine and the Pursuit of Health', *Public Interest* **40**, 11–42.
19. Konald, D.: 1962, *A History of American Medical Ethics, 1847–1912*, The State Historical Society of Wisconsin, Madison.
20. MacIntyre, A.: 1981, *After Virtue*, Notre Dame Press, Notre Dame, Indiana.
21. MacIntyre, A.: 1978, 'Historical Perspectives on the Ethical Dimensions of the Patient-Physician Relationship: The Medical Ethics of Dr. John Gregory', *Ethics in Science and Medicine* **5**, 47–53.
22. McCullough, L.: 1981, 'Justice and Health Care: Historical Perspectives and Precedents', in E. Shelp (ed.), *Justice and Health Care*, D. Reidel, Dordrecht, Holland. pp. 37–50.
23. McCullough, L. and Beauchamp, T.: 1983, *Medical Ethics*, Prentice-Hall, Englewood Cliffs, New Jersey, esp. Chapter Two.
24. Nelson, L.: 1978, '*Primum Utilis Esse*: The Primacy of Usefulness in Medicine', *The Yale Journal of Biology and Medicine* **51**, 655–677.
25. Percival, T.: 1803, *Medical Ethics, or a Code of Institutes and Precepts Adapted to the Professional Conduct of Physicians and Surgeons*, S. Russell, Manchester.
26. Picher, L.: 1883, *An Ethical Symposium: Being a Series of Papers Concerning Medical Ethics and Medical Etiquette from the Liberal Standpoint*, G. P. Putnam's Sons, New York.
27. Veatch, R. M.: 1981, *A Theory of Medical Ethics*, Basic Books, New York.
28. Wills, G.: 1978, *Inventing America: Jefferson's Declaration of Independence*, Doubleday, Garden City, New York.
29. Wittgenstein, L.: 1953, *Philosophical Investigations*, Basil Blackwell, Oxford.

JOHN DUFFY

AMERICAN MEDICAL ETHICS AND THE PHYSICIAN-PATIENT RELATIONSHIP

The most difficult tasks in determining the effect of medical ethics on the doctor-patient relationship are first to separate ethics from economics and second to differentiate between medical etiquette and medical ethics. Prior to the twentieth century most American physicians were poorly paid and whatever respect and prestige they accrued depended largely upon personal factors. Hence the early efforts of physicians to achieve professional unity were motivated by economic considerations. The aims of all early medical societies were to reduce competition from irregular practitioners, establish uniform fee bills, and eliminate public quarrels over medical theories and practices among the orthodox physicians. While the many attempts by local and state societies to establish ethical codes prior to the American Medical Association's code in 1847 affected doctor-patient relationships both favorably and unfavorably, ethical codes were only one of several factors bearing on this relationship.

By the end of the colonial period at least three classes of physicians had emerged: the highest level included those who held college degrees, had served an apprenticeship, and had supplemented their education by study in Great Britain and/or on the Continent; a second category, the largest one, consisted of men trained by the apprenticeship system who may or may not have continued their education by further reading; the lowest group were the empirics, bone-setters, herbalists, and other self-appointed practitioners ([8], p. 507). While the leading physicians enjoyed prestige and social status, primarily owing to their position as members of the upper class, the majority of practitioners struggled to make a living. Entrance into the profession, if it can be called that, was easy, and the resulting competition for patients prevented any sort of professional unity. The state of medical knowledge was such that few doctors could agree on either medical theory or practice, and dissension characterized even those well-to-do doctors whose economic position was secure. The early medical societies rarely survived for long, since disputes over theories, therapeutics, and practices usually degenerated into personal quarrels, punctuated by public denunciations of each other's therapeutic measures in newspapers and pamphlets ([16], p. 32).

The eighteenth century was predominantly an age of medical theories, one

Earl E. Shelp (ed.), The Clinical Encounter, 65–85.

in which physicians were desperately searching for the one universal law of health. Faith in the traditional humoral theory was steadily declining, but none of the new theories or modifications of the humoral concept proved satisfactory. Scientific knowledge was growing, but the human organism is exceedingly complex and advances had to be made on a broad front before medicine could benefit. In the eighteenth and early nineteenth centuries physicians still had no understanding of infectious diseases and very little conception of the nature of constitutional and organic disorders. In despair they intensified their application of the traditional remedies of bleeding, blistering, purging, vomiting, and sweating, and prescribed horrendous dosages of mercurials, arsenicals, and other deadly drugs. The result was a widespread distrust of the medical profession, a suspicion only strengthened by the bitter public quarrels among physicians. Yet, despite public criticism of the profession, individual physicians were held in high respect, an ambivalence which holds true today.

Recognizing the need for professional unity, medical practitioners organized a number of local and state medical societies in the late eighteenth century. The main purpose of these organizations was to encourage an esprit de corps, promote professional ethics, improve the members' incomes, and combat the growing number of quacks and irregulars ([38], pp. 20–21; [28], p. 136). Coincidentally in 1794 an English physician, Dr. Thomas Percival, wrote a book on medical ethics as a guide for his son who intended to enter medicine. The book, entitled *Medical Ethics*, was published in Manchester, England, in 1803 and became the basis for virtually all American ethical codes [4]. Well before this time, groups of physicians were drawing together for their mutual benefit. Almost invariably the first step in this direction was the issuance of fee bills. While efforts to achieve professional harmony could only serve to improve doctor-patient relations, the same could not be said of fee bills.

Ths first medical association organized on a colony-wide basis was the New Jersey Medical Society founded in 1766. The Society's original minute-book speaks of the "low state of Medicine in New Jersey" and "the many difficulties and discouragements, alike injurious to the people and the physician." The assembled physicians were meeting together "for their mutual improvement, the advancement of the profession and the promotion of the public good." At this first meeting in July, 1766, officers were elected, a constitution established, and then the physicians turned to the real purpose of the gathering, to draw up a fee bill ([14], pp. 10–12).

The fee bill itself was long and detailed and carried a preamble intended to allay the suspicions of prospective patients. It cited the long years of training required for physicians, the dangerous and disagreeable tasks associated with medicine, and asserted that standardizing fees would eliminate disputes between physicians and patients. It warned that members who reduced their fees except for certain "laudable motives" were to be subject to expulsion from the Society. As was to happen to a great many fee bills in the succeeding years, a public outcry forced the Society within four months to rescind the measure. According to the minutes: "Evil-minded persons had thrown an odium" on the Society's proceedings and prejudiced "the minds of the inhabitants" against the physicians. Another forty years elapsed before the Society issued a second fee bill ([14], pp. 10–12; [18], p. 318).

The problem of medical fees was a sensitive one for both doctors and patients. Physicians who encouraged the assumption that their role was one of dedication and service, part priest and part parent, could scarcely be businessmen at the same time. In the colonial period and well into the nineteenth century medical bills were permitted to drag on for years, in some cases until the estate of the patient was settled. As medical bills mounted, collection became more difficult, and unpaid bills were the perennial bane of physicians. One provision in nearly all fee bills called for a regular settlement of accounts and recommended that they be closed at the end of each year ([33], pp. 6–7). One reason for the financial problems of the medical profession was the long American tradition of do-it-yourself, a tradition that throughout our history has applied equally to medicine. Unaccustomed to paying for medical care and always short of cash, Americans generally were reluctant to pay medical fees. Money was scarce, particularly among the lower income groups, and free care for the poor was an accepted practice among physicians. An article in the *North American Review* in 1831 stated that although there was "little abject poverty" in Boston, "there are some thousands, who never dream of such a thing as a physician's fee. . . ." "The custom of attending the poor gratuitously," the article continued, "has now grown, by long usage, into a sort of common law. . . ." ([1], p. 379). In 1855 the *Boston Medical and Surgical Journal* advised doctors not to charge more than $2.00 for a visit on the grounds that the public would not pay more ([26], pp. 56–57). The situation in rural areas was at least as bad. A late 19th century practitioner in upstate New York lamented that his patients objected to paying more than fifty cents for a home call and more than thirty-five cents for an office visit, and added that he often had to wait one to five years to receive his fee ([33], pp. 88–89).

The life of most of the fee bills promulgated in the nineteenth century was short. In the first place medical associations were rarely able to enforce the provisions, and too often internal quarrels within the organization or fierce competition from non-members negated efforts to standardize medical fees. A good many outstanding physicians whose positions were secure felt no need for them. Benjamin Rush in 1789 lordly advised his colleagues to adapt the fee to the circumstances of the case and the patient ([37], pp. 155–56). By this date Rush was probably the best known physician in the country, and he must have had little appreciation of what the average practitioner was facing. The latter was competing with an excessive number of his own colleagues and an equally large number of irregular practitioners for a very limited number of patients who could afford medical services, a situation which was to remain true until the American Medical Association successfully limited entrance to the field of medicine in the twentieth century. A fee bill put forth by the doctors of New York City in 1790 was reissued in 1798, 1816, and again in 1825. In connection with its publication in the latter year, a writer in the *New York Monthly Chronicle of Medicine and Surgery* complained of the bitter quarrels among physicians and commented that "amongst the most prominent of the sources of discord, is the subject of *fees*" ([33], p. 2).

Of equal importance with dissension among physicians in negating the effort for standardized fees and professional esprit de corps was the egalitarian spirit which characterized the early nineteenth century, the Age of Jacksonian Democracy. The prevailing attitude was that any good American could do any job – whether it involved carpentry, government, law, religion or medicine [32]. Along with it came a strong anti-intellectualism which decried 'book-learning', an attitude scarcely consonant with the aims of the better physicians to improve medical training and raise the intellectual and ethical standards of the profession. Even those laymen who did not question the harsh therapy of orthodox physicians shared the common suspicion of monopolies, a word in the nineteenth century that had all the connotations of the term 'communism' today. Fee bills were taken as clear evidence that physicians were attempting to create a monopoly, and public reaction, while it varied from place to place, was often quite sharp. For example, when a group of doctors in the District of Columbia issued a fee bill in 1833, a series of public protest meetings were held during which it was suggested that new doctors be invited into the District ([24], pp. 90–91).

The early codes promulgated by medical societies, as noted earlier, sought to prevent quacks and irregulars from practicing. A number of societies successfully advocated state licensure laws and in some cases obtained the power to control medical licensing. In addition, they forbade their members to consult with irregulars. These early successes in securing state licensure laws proved not only meaningless but downright harmful. In the first place the laws were weak, and there was scarcely a judge or jury willing to convict a quack or irregular physician for practicing without a medical license. Moreover, it seemed to confirm lay opinion that physicians aimed to monopolize medical practice ([38], p. 27; [15], p. 6). The enactment of licensure laws coincided with a rising distrust of orthodox medicine engendered by what has been termed the heroic age of medicine. The public reaction to the drastic bloodletting, purging, and vomiting of the regulars created an atmosphere favorable to the more moderate therapies of homeopaths, botanical doctors, and hydropaths. These irregulars, by leaving much of the cure to nature, were often more successful than those orthodox physicians who rigorously practiced traditional methods.

The one group of irregulars who most threatened orthodox physicians was the homeopaths. Founded by Samuel Hahnemann in the late eighteenth century, homeopathy was introduced into America in the 1820s and made rapid gains. Its success was based to a large extent upon one of its basic principles that the more minute the dosage the more efficacious the remedy. Its practitioners, operating on this principle, diluted their drugs to such a degree that they practiced therapeutic nihilism. Homeopathy threatened the regular profession in two ways. Most of its early converts came from among the regulars, for a good many physicians quickly recognized that homeopathic treatment, for whatever reason, was more effective than the traditional one. In the second place, Hahnemann had emphasized that disease resulted from spiritual and dynamic disturbances ([25], p. 134). The 1830s and 1840s was an age when the American middle class was espousing transcendentalism and spiritualism, and homeopathy, which provided mild treatment and an appealing rationale, began making gains among the well-to-do patients who formed the basis of a successful medical practice. The hydropaths, too, who usually treated patients in their 'Institutes', also appealed primarily to this same group. On the other hand, the Thomsonians, herb doctors, and other empirics had their greatest appeal in rural areas and among the lower income groups. Thus by striving to suppress irregular practitioners, the medical societies antagonized a broad spectrum of society ([16], pp. 109–123).

To summarize the situation, by the 1830s irregular practitioners were making large gains at the expense of the orthodox profession, and public opinion was tending towards egalitarianism and anti-intellectualism. Under these circumstances, efforts by physicians to establish professional unity and raise standards could scarcely have come at a worse time. The widespread suspicion of the orthodox medical profession is reflected in what was essentially a sympathetic comment on the rules and regulations of the Boston Medical Association in the *North American Review* in 1831. The author wrote that the profession was held in "low repute" in part because of "a bigoted attachment to authorized modes of practice and a peculiar sensitiveness of feeling, or readiness to take offense for slight causes" ([1], p. 368). The comments of physicians and the actions of state legislatures are even more revealing. Nicholas Romayne, president of the New York State Medical Society, stated in 1810 that since the public was so opposed to licensure laws, the only way to eliminate quackery was for regular physicians to demonstrate the superiority of their therapy. Unfortunately, the regulars were in no position to do this in the early nineteenth century. A few years later, when Congress granted a charter to the Medical Society of the District of Columbia, it empowered the society to license members, but it specifically denied them the authority to issue a fee bill or code of ethics ([27], p. 30).

During the 1830s and 1840s the strong public reaction against monopolies in general and the medical profession in particular led to efforts to rescind nearly all state medical licensure laws. A committee appointed by the Louisiana State Medical Society to consider medical education and the state's medical license law reported in 1851 that the existing law was useless. In recommending that the law be strengthened, the committee members saw little hope since they felt the public would never "consent to the enactment of a penal statute for the protection of this or any other science." The report added in a tone of utter discouragement that laymen cared "little for the intrinsic merits or qualifications of a man . . . for they think that . . . it is quite as easy to cultivate cabbage as science." This attitude was so widespread that many medical societies during these years were reluctant to support medical licensure laws on the grounds that any such action would only increase public sympathy for the irregulars ([19], pp. 113–14; [35], p. 78).

By the 1840s the American medical profession was at a low point in its history. The early state licensure laws had been swept away or rendered meaningless by the enactment of subsequent laws. For example, Maryland had passed one of the better licensure laws in 1799 requiring the licensing of physicians and levying a $50 penalty for practicing without a license. While not strictly enforced, it did provide a means for differentiating between orthodox and irregular practitioners. In 1839, responding to the egalitarian movement, the Maryland legislature authorized Thomsonians and botanic physicians to charge for their services. This law in effect negated the licensure law. Following its passage, few physicians bothered to apply for licenses, and the state medical society received a major setback ([13], pp. 102–105). With the profession torn by internal dissension and a wide variety of medical sects making inroads into the patient population, a group of physicians representing the leading medical societies and the better medical schools came together in an effort to establish a national medical association.

In May of 1846 an organizational meeting was held in New York City, and one of the first resolutions declared "that it is expedient that the medical profession in the United States should be governed by the same code of medical ethics" and directed that a committee of seven be appointed to report on the subject at the next meeting. The following year at a second meeting in Philadelphia, the American Medical Association came into formal existence, and an ethical code, based upon Percival's *Medical Ethics*, was adopted ([11], pp. 1–2). In speaking in favor of the code, Dr. John Bell, a member of the ethics committee, insisted that physicians should be entitled "to, at least, the same respectful and considerate attentions that are paid . . . to the clergyman, . . . and to the lawyer . . . ." Physicians by the nature of their service, he continued, were in the best position "to exhibit the close connection between hygiene and morals . . . ." He deplored the fact that many newspapers and clergymen were aiding and abetting "the enormities of quackery," and he criticized apothecaries for in effect allying themselves "with empirics of every grade and degree of pretention [sic]" ([31], pp. 26–27).

The first part of the code dealt with relations between physicians and their patients. It stated that a patient should expect to receive from his physician close attention, humanity, confidentiality, frequent visits, proper balance between hope and truth, comfort when dying, consultation in a difficult case, and moral guidance. The duties of the patient in turn were to select a trained orthodox physician and consult him on even trivial medical complaints, communicate fully with him and make him "his friend and advisor," obey medical prescriptions and advice on all matters, avoid even friendly visits of a physician not in attendance, "never send for a consulting physician without the express consent of his own medical attendant," and have "a just and enduring sense of the value of the services rendered him by his physician" ([36], p. 6).

The next section of the code, and the longest one, was in effect a code of medical etiquette rather than ethics, since it specified the duties of physicians to each other and to the profession at large. For most of history physicians, lacking any real understanding of illness and disease, have been compelled to rely upon their moral authority in treating patients. Hence it is not surprising that the second paragraph of this section began with the statement: "There is no profession, from the members of which greater purity of character, and a higher standard of moral excellence are required, than the medical," a statement which ministers and other professionals might take amiss ([36], p. 7). In 1848 it certainly represented an ideal rather than actuality, for President Nathaniel Chapman in opening the first annual meeting of the American Medical Association in Baltimore that year explained the need for a code by conceding: "The profession to which we belong, once venerated on account of its antiquity – its various and profound science – its elegant literature – its polite accomplishments – its virtues – has become corrupt and degenerate to the forfeiture of its social position and, with it, of the homage it formerly received . . . ." ([20], p. 41). Since medicine originated and long remained closely associated with religion, its early practitioners received the respect due to priests and medicine men. With the secularization of medicine, however, its status has varied considerably, but until well into the twentieth century physicians were harking back to halcyon days that had never existed on the American scene.

The code was completed by a section treating the duties of the profession to the public and those of the public to the profession. The entire code was quickly accepted by medical societies throughout the country, and as new ones were formed, they, too, subscribed to it. Occasional disputes over its acceptance occurred, but the desperate state of the profession made the need for professional unity all too apparent. There were, of course, a few rugged individuals who had little use for associations of any kind. President Daniel Brainard of Chicago's Rush Medical College, an outstanding medical figure, considered medical societies mere trade unions designed to raise fees, or "punitive leagues" for the purpose of enforcing professional ethics ([9], p. 73).

A good many other physicians resented attempts to enforce the code, considering them an effort to form a monopoly or a restriction on personal freedom. As late as 1883 Dr. Lewis S. Pilcher, in a symposium on medical ethics and etiquette, sharply attacked the AMA's code of ethics on liberal grounds. While conceding the nobility of the medical profession, he proclaimed: "Every principle and instinct of manhood leads an individual to assert his right of independent judgment in matters that pertain to his feelings and conduct. . . ." He emphatically denied a claim made that year by Dr. John L. Atlee, president of the AMA, that the code had been of great benefit to physicians. In Pilcher's opinion the code had fostered "a spirit of censoriousness in the profession" and made "every man a spy upon his neighbor." Most of the code's provisions, he continued, had been ignored, and all attention had been concentrated upon the single issue of the irregulars "in such a way as to create for it sympathy and to promote its growth in the esteem of the public" ([31], p. 36).

Public fears, and those expressed by individual physicians, that the code would immediately create a medical monopoly were needless. The AMA and its associated societies represented only a fraction of American practitioners, and their membership grew slowly, a situation which hampered enforcement of the code. The two major weapons medical societies could bring to bear were social pressure and the threat of expulsion. But until a fairly high percentage of physicians belonged to the local medical society, these tactics had only limited value.

The one section of the code which created the greatest difficulty within societies and which most affected doctor-patient relations was the section prohibiting members from consulting with irregular practitioners ([35], pp. 170–74). Unfortunately many of these irregulars, particularly homeopaths and eclectics, had gained a considerable measure of respectability in their communities and even in the eyes of many orthodox physicians. Adherence to the code in some instances meant breaking close personal and professional ties. In small communities and neighborhoods supporting only two or three physicians, refusal to consult with a homeopath or other irregular often denied the patient any chance for consultation. To complicate matters further, in a period when medical knowledge was acquired by the apprenticeship system and a nominal course of medical lectures, the line between regulars and irregulars was a fine one indeed. The case of Dr. Morton Robinson of Newark, New Jersey, illustrates this point quite well. Robinson began practicing as a botanic physician around 1851, probably on the basis of an apprenticeship, since he did not graduate from a botanical medical school until 1854. He was listed in the *City Directory* as a botanic physician until 1871. In 1872 he was included under the heading 'allopath', a term applied to regular physicians. In 1877 he was president of the Eclectic Medical Society of the State of New Jersey, but from 1883 to 1894 the *Directory* simply classified him as 'physician and surgeon' ([14], p. 72).

The career of Dr. Robinson is not unusual, since a good many physicians in the early nineteenth century made the transition from regular to homeopath or eclectic, and later, as homeopathic and eclectic medical schools emerged, their graduates frequently moved over into orthodox medicine. To make the situation even more confused, orthodox medicine was itself in a state of flux. It had discarded both the traditional humoral concept and the newer theories of the seventeenth and eighteenth centuries, but for much of the nineteenth century neither clinical medicine nor advances in science supplied it with a satisfactory substitute.

In consequence throughout the rest of the century internal dissension over the definition of irregulars and the question of whether or not to consult with them continued to hamper the activities of the American Medical Association and its constituent societies. In 1870 two physician members of the Boston Gynecological Society protested to the AMA that the delegates from the Massachusetts Medical Society should not be seated since the Society had admitted homeopaths and eclectics to its ranks. When its delegates were rejected, the Massachusetts State Society refused to send any representatives to the next annual meeting, although it did submit a long report to the secretary of the AMA defending its orthodoxy. Although it created bitter feeling between the Boston Gynecological Society and the Massachusetts Medical Society, the affair stimulated the latter body to start proceedings against members who espoused homeopathic or eclectic medicine ([10], pp. 127–31, 425–27).

Purists in the medical societies objected to their members having any connection with irregulars. In 1872 the American Medical Association criticized some of the hospitals and medical schools in Washington, D. C., because they contained unlicensed physicians on their staffs. The issue was still in doubt in 1878 when the Michigan State Medical Society sent Dr. Edward Dunster of the University of Michigan Medical School faculty as a delegate to the AMA. Although he was an orthodox physician, the Michigan faculty included two homeopathic instructors, and Dunster was accused of "aiding and abetting the graduation" of irregulars. As a result of the ensuing furor, which was not finally settled until the 1881 AMA meeting in Richmond, an amendment to the code proscribed members from signing diplomas or certificates for persons supporting or intending to practice irregular medicine ([20], pp. 79, 84; [24], pp. 107–109).

The bitterest quarrel within the AMA over the issue of irregulars surfaced in 1882 and involved the New York State Society. The problem originated in 1881 when the Medical Society of New York County admitted two homeopathic medical school graduates who had subsequently turned to orthodoxy. At this time a movement was gaining strength in the New York State Society to relax the prohibition against consulting with irregulars. In consequence, in 1882 the Society's code of ethics was revised to permit consultation with homeopaths. The debate over the change was heated and bitter, and it split

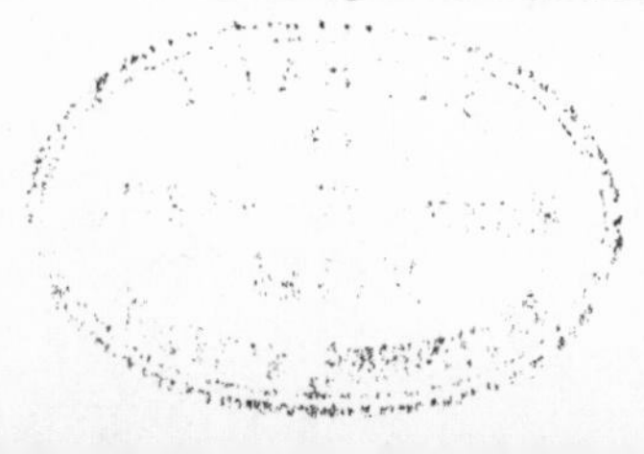

the State Society. When the AMA refused to seat the New York delegation in 1882, the minority faction which opposed the alteration in the state ethical code broke away and established a new association. Thus for the next twenty years at the state, county, and municipal levels, two complete medical societies existed side by side in New York State ([35], pp. 302–305).

The major argument put forth by those who favored consulting with homeopaths was that the existing proscription in the code merely helped them. Dr. Henry Piffard, a leading physician, wrote in the *New York Medical Journal* that "the general effect of the 'code' was . . . to build up and strengthen the sectarian societies . . . by exciting public sympathy in their favor, and thus aiding them politically" ([35], p. 302). Another telling argument was that the relatively high percentage of homeopaths in urban areas meant that they had strong public support. In New York State, as elsewhere, medical societies were pressing for licensure laws but with only limited success. Besides a basic suspicion of monopolies or 'trusts', patients of homeopathic and eclectic physicians resented the attacks on their doctors, and they opposed all licensure laws sponsored by medical societies. In this opposition they were joined by many voters who believed in the right of any individual to choose his own form of medicine.

As the nineteenth century drew to a close, increasing medical knowledge and better education steadily improved the public image of the medical profession, and the movement for medical licensure gradually began to make headway. The two chief obstacles to passage of these laws, as just noted, were the opposition of the irregulars, led by the homeopaths and eclectics, and the deep-rooted public belief in individual rights. The homeopaths, osteopaths, and other irregulars had no illusions as to what would happen to them were the licensure laws proposed by the orthodox medical societies to be enacted, and they mobilized their patients and supporters in firm opposition to those measures.

Before the battle for medical licensure could be won, a major change had to occur in public opinion. As those opposed to the AMA consultation provision had argued, its net effect was to strengthen the irregulars. Laymen viewed physicians' arguments over medical theories in the same light as theological disputes, and they saw no reason for giving the proponents of any particular theory a monopoly. In disputes between regulars and homeopaths, the newspapers were generally supportive of the latter. When the Medical Society of the District of Columbia charged Dr. Christopher C. Cox with breach of the ethical code for serving on a board of health along with a homeopath, the New York *Times* scornfully condemned the action.

Commenting on the regular profession's attitude towards homeopaths, the *Times* editor wrote: "There is no stronger tenet in the orthodox creed than that it is better the patient should die under the old remedies than recover under homeopathic treatment" ([24], p. 87).

The national magazines, too, viewed the arguments between medical sects with a mixture of amusement and derision. When the AMA expelled the New York delegates because their medical society had agreed to consult with homeopaths, the *Nation*, in an editorial entitled 'The Medical War', suggested that if physicians were so sure the homeopaths were wrong they should be willing to consult with them for the benefit of the patient [3]. Some ten years later, in 1893, a quarrel broke out over the proposed appointment of a homeopathic physician to a hospital in Westchester, New York. The editor of *Harper's Weekly* treated the whole issue as a tempest in a teapot, declaring with tongue in cheek that "there is only one thing that is harder to get out of a hospital than an allopathic doctor, and that of course is his patient." He continued cheerfully that "in the good time coming, the homeopath and the allopath will lie down together, and the Christian Scientist will treat them," adding, however, that he felt this was some years away [2].

Between 1880 and 1900 the social climate with respect to medicine began to change. The worst abuses in traditional medicine, such as drastic bloodletting and excessive drugging, had given way to more moderate practices, and major advances in all the sciences were beginning to provide medicine with a more satisfactory basis. The quality of American education was improving, and this held especially true for medical schools. As more emphasis was placed on the basic sciences and clinical training, the gap between homeopaths and regulars steadily narrowed. In addition to these factors, although medical societies generally supported the AMA code of ethics, efforts to enforce the consultation provision were considered by many members to be a form of persecution. Increasingly the question was asked as to whether or not one could force physicians to become moral.

At the same time public opinion was moving in two opposing directions. The Progressive Movement at the end of the century revived the old fears of monopoly, and the Sherman Antitrust Act of 1890 and subsequent antitrust laws were a direct threat to the AMA. Moreover, the American faith in individual liberty was a constant factor. In the 1890s a National Constitutional Liberty League was founded in Boston to lobby for freedom of choice in medicine, and the League had many sympathizers. One of these, Samuel Clemens, was reported as saying: "I don't know that I cared much about these osteopaths until I heard you were going to drive them out of the State;

but since I heard this I haven't been able to sleep . . . ." Clemens added that his body was his own, not the state's, and that he had the right to do with it what he wished ([38], pp. 59–60).

On the other hand the complexity of society was forcing an increasing degree of professionalization, and there was a growing recognition that physicians, above all professionals, needed special credentials, since their services involved life and death. An article in the *New Englander and Yale Review* in 1889 asked rhetorically whether or not it was possible to awaken "the people of the United States to the fact that the medical profession holds the lives of men, women and children in the hollow of its hands," and, in view of this, why there was no law requiring the members to at least "be technically educated" ([39], p. 134). Twelve years later a series of articles in the *Forum* praised the AMA for its stand against medical sects. In one of these, Champe S. Andrews commended "the great body of the profession" for its battle against illegal medical practice, and in another P. Maxwell Foshay argued that with the new scientific medicine only the ignorant or immoral would advocate one cure for all diseases ([7], p. 542; [21], p. 167).

Aside from qualms about the rigidity of the consultation provision and the possible threat arising from the anti-trust laws, even the staunchest supporters of the AMA code recognized that without the cooperation of homeopaths and eclectics there was little chance for passing medical licensure laws. Further, as mentioned earlier, the advance of science had narrowed the gap between the better sectarians and orthodox physicians, thus making it easier for the AMA to bow to the inevitable. In state after state orthodox medical societies collaborated with sectarian societies to establish licensing boards. The form of the licensing agencies varied but generally there was either one board on which homeopaths and eclectics served or else the latter were given separate boards. By 1898 only the Alaska Territory had no regulations concerning the practice of medicine. The western states lagged behind in enacting licensure laws, and it was this situation that made the Kansas Medical Society take the unusual step of agreeing to meet jointly with the homeopathic and eclectic societies in 1898 and 1899 ([38], pp. 54–55; [22], pp. 52–57).

By 1900 homeopaths, eclectics, and certain other medical sects had achieved legal status by virtue of the licensure laws, a situation which made the AMA's prohibition against consultation with them difficult to enforce. In addition, the regular profession, bolstered by advances in medical knowledge, was feeling more confident of its position and hence could afford to be more tolerant. In 1902 AMA President John Allen Wyeth, in urging that the consultation provision be repealed, clearly reflected the sentiment of the membership. Liberal physicians, who had long felt the code was too detailed and rigid, used this opportunity to rewrite it. The consultation provision was replaced by a simple statement: "The broadest dictates of humanity should be obeyed by physicians whenever and wherever their services are needed to meet the emergencies of disease or accident." An equally important change was the omission of the long section detailing the various ways in which the patient should be obedient and loyal to his physician ([5], pp. 4–6). This section was worse than useless since the vast majority of patients had never read it, and those who did reacted with irritation.

A sharp personal attack on the Secretary-General of the AMA in 1909 resulted in a move to strengthen the enforcement provisions of the code of ethics, and a new 'Principles of Medical Ethics' was adopted in 1912. No major changes were made in the code, and this version, modified to meet changes in twentieth century medicine, remained in effect until 1957. At that time the entire code was replaced by an abridged 'Principles of Medical Ethics', consisting of a brief preamble and ten short sections ([26], pp. 70–74).

The inextricable relationship between ethical codes and economics noted earlier is shown again in the enduring efforts of the AMA to protect the fee system. During late eighteenth and early nineteenth century dispensaries appeared; these were outpatient clinics established in part from philanthropic motives and in part by the need of medical schools to provide clinical training for their students. A good many were also established by specialists wishing to improve their techniques for dealing with particular types of cases. The dispensaries provided medical care to the poor either gratuitously or for a nominal charge. Usually they were financed by private individuals or organizations, but most of them received some support from municipal or state governments. As medical knowledge and education advanced in the late nineteenth century, the services of physicians increasingly became restricted to the middle and upper classes. The poor and unemployed were forced to rely upon dispensaries and clinics, and even the better paid working class resorted to private physicians only as a last resort. In consequence the dispensary movement became a major factor in health care. In New York City, for example, twenty-six dispensaries in 1871 treated 219,851 patients, about twenty per cent of the city's population. By 1893 the number of dispensaries had grown to sixty-four, and in 1900 almost 900,000 patients were treated in them ([17], p. 185; [34], p. 33).

The average physician, as noted earlier, was always beset with financial problems, and as early as 1852 a series of resolutions were introduced in the New York Academy of Medicine denouncing the free care provided by dispensaries. One of them called such charity degrading to the medical profession, while another declared that all gratuitous medical help savored of "quackery" ([29], pp. 288–89). Although the resolutions were tabled, in the succeeding years these denunciations grew in direct proportion to the growing number of dispensaries. Just when medicine was beginning to achieve a measure of respectability, a host of new medical schools appeared on the scene. These institutions simultaneously established new dispensaries and began graduating a flood of new physicians; hence while the prestige of the better physicians was improving, the average doctor found his economic position deteriorating. The major charge against the dispensaries was that they provided free medical care indiscriminately. Most physicians had no objection to medical care for the so-called 'deserving poor', but they felt that many dispensary patients could pay the relatively low fees charged by private doctors.

Fortunately for physicians, advances in medical knowledge and practice signaled the end of the dispensary in the early twentieth century. It had functioned fairly effectively when all a physician needed was a stethoscope for diagnosis and a few simple drugs. With the rise of laboratory medicine, clinical training moved into hospitals, and low-cost institutions such as dispensaries simply could not provide adequate medical care ([34], pp. 49–51). The rising standard of living meant, too, that the working class could more easily afford private medical care. Whether or not this change affected doctor-patient relations is hard to say. Dispensary patients represented the lowest income group, one that has always been largely inarticulate. Those patients whose upward mobility enabled them to employ private physicians generally accepted the American ideal of private enterprise and were not likely to complain.

While opposition to the dispensaries did not directly involve the AMA ethical code, the contract practice of medicine has always been of concern to the AMA and its Judicial Council. Contract medicine took three major forms in the nineteenth century. Southern planters frequently employed physicians on an annual basis to provide medical care for everyone on the plantation. In the North, particularly in urban areas, social, fraternal, and volunteer groups and trade unions frequently contracted with a physician to care for all members for a year on a fixed per capita basis. A third area where contract medicine was making rapid gains was in industry, particularly in mining, lumbering, and railroads. In addition, by the early twentieth century hospitals began offering medical coverage for entire families.

The average physician in these years was barely surviving economically, leading to fierce competition for contracts. For many young physicians without established practices, the contract method, despite its miserable pay, was the only alternative. Medical societies reacted with considerable ambivalence to contract medicine, although nearly all condemned the low pay. To make matters worse, physicians soon discovered that fraternal orders, insurance companies, and other middlemen were keeping the major share of fees collected from members. In the early twentieth century, the elimination of many low grade medical schools and a corresponding reduction in the physician-to-population ratio encouraged medical societies to begin tackling the contract system. In 1911 a committee of the Erie County Medical Society of New York charged that contract medicine was "unethical, unjust, and in every sense injurious to the profession in general" ([12], p. 131). The following year the House of Delegates of the California State Society resolved that its members could not accept contract work for any families with an income of $75 a month or more ([12], p. 128). Up to this time the AMA had taken no stand on the issue. In 1913 its Judicial Council reported that since three-quarters of American workers had incomes of less than $600, medical insurance was necessary, but it did suggest that some type of income limitation should be established ([30], p. 1997). Its constituent societies, however, were beginning to take a firmer stand, with some of them demanding that physicians who practiced contract medicine be expelled.

During World War I the major issue in the health field was the fight for compulsory state medical insurance. Organized medicine successfully led the battle against it, and by so doing firmly established the fee-for-service system of private medicine in the United States. The contract issue was not revived until the Great Depression when hospitals, in desperate financial straits, reinstituted hospital insurance, and the federal government began moving into the health area. In their fight against what it termed 'state' or 'socialized' medicine, the medical profession argued that any form of medical practice other than the fee-for-service system was unethical, since it interfered with the personal relationship between doctor and patient, restricted patient's choice of physician, and limited the physician's control over treatment. Although the original AMA code and the revised version of 1903 did not specifically forbid contract medicine, the Association and its constituent societies, professedly on ethical grounds, fought bitterly against every innovation in health care distribution from Blue Cross to Medicare and Medicaid. What is surprising is that this opposition had so little effect on doctor-patient relations or even on the general image of the medical profession. The probable explanation is the steadily rising standard of living after 1935 and the major medical breakthroughs which so enhanced the status of the medical profession in the postwar years. Buoyed up by the public's unbounding faith in science, the image of the American medical profession probably reached its zenith around 1960. Since then the phenomenal rise in medical costs, the impersonality of modern medicine, and other factors have caused the profession to lose a good share of its lustre.

One of the most sensitive areas in the medical profession relates to malpractice. The first AMA code of ethics, while emphasizing the need to fight charlatans and unorthodox practitioners, said nothing about exposing the errors and ignorance of its own members. The entire emphasis was on keeping all disagreements among members strictly within the profession. The code specified that when controversy and "contention" occurred, the matter had to be decided by a "court-medical", adding that insofar as the public was concerned "a peculiar reserve must be maintained by physicians . . . in regard to professional matters . . .." Since medical concerns "cannot be understood or appreciated by general society . . . ," the "adjudication of the arbitors" should not be made public, as "publicity in a case of this nature may be personally injurious to the individuals concerned, and can hardly fail to bring discredit on the faculty" ([36], p. 12). Even as late as 1903, the code spoke only of an obligation to expose "the great wrongs committed by charlatans," but said nothing about those committed by members of the profession ([5], p. 6). It was not until the 1912 revision that the code specifically mentioned a responsibility to disclose "before proper medical or legal tribunals, corrupt or dishonest conduct of members of the profession" ([6], p. 62). The physician was also expected to prevent the admission of unfit or unqualified members, but nothing was said of those already in the profession.

Since medicine is still an art utilizing scientific knowledge and methodology, much of what physicians do even today represents value judgments, and it is precisely this factor which makes practitioners reluctant to criticize each other. In addition, confidence is of vital importance in the doctor-patient relationship. The AMA was justified in endeavoring to reduce the notorious discord within its ranks, but, as is often the case, the pendulum swung too far. As the profession closed ranks, physicians sought to make medicine even more abstruse by utilizing technical jargon and insisting on rigid obedience to their prescriptions. Medical societies frequently discouraged their members from testifying against each other. For example, the Michigan Medical Society in 1887 ruled that it was unethical for any member to testify against another member in malpractice cases. By 1900 there was a general tendency for physicians to conceal each other's mistakes, misjudgments, and often outright blunders on the grounds that this was necessary to maintain "the general respect of the profession . . . ." ([26], p. 49).

By the mid-twentieth century physicians had assumed the mantles of both priest and scientist, and the secrecy which surrounded the doctor-patient relationship and the profession's image was at its peak. As noted earlier, the rising cost of medicine and its growing impersonality in the succeeding years weakened the public image of the profession and contributed to the growing number of malpractice cases. Ironically, laboratory and hospital medicine, by introducing more exactitude into medicine, can take some credit for legitimate malpractice cases, although the rapidly growing number of lawyers, ethically dedicated to the legal fee system, probably accounts for the majority of cases.

One area in which medical codes of ethics have had little to say concerns the question of animal and human experimentation. Without such experimentation, progress in medicine would have been sharply limited, yet the possibilities for abuse are grave. Until quite recently, social misfits, prisoners, the poor, and minority groups have been quite acceptable for experimental purposes. The cesarean section, which was considered almost a hopeless operation until well into the nineteenth century, is a fine illustration of this point. In the 1820s a Louisiana physician, François Prevost, performed at least four of these operations upon slave women, and between 1822 and 1861 some eleven more cesarean sections were done on slave women. It is a commentary upon the times rather than the physicians that at no time in Louisiana was this operation attempted upon a white woman.

The more recent Tuskegee syphilis experiment, described by James H. Jones in his book, *Bad Blood*, also illustrates that ethical codes apply only to the typical or respectable members of society [23]. The four hundred experimental subjects in this case had everything against them; they were poor, ignorant, Black, and suffering from syphilis, a disease traditionally associated with immorality. The treatment for syphilis at that time was long, painful, and of limited value, and it is doubtful that if the public and general medical profession had known of the experiment there would have been much of an outcry. The real moral failure, aside from the scientific inadequacy of the experiment, was the refusal by the physicians associated with it to end the experiment in the 1950s when penicillin was proving a miracle drug in the cure of syphilis and other infections. The outrage which followed the exposure of this on-going experiment in 1972 reflected the changed attitude of the public and the medical profession, and in consequence led to more stringent regulations with respect to human experimentation. While the medical profession undoubtedly lost some prestige as a result of this exposé, the blame was placed largely upon medical scientists rather than practitioners, and whether or not this event had any serious impact upon doctor-patient relations is doubtful.

The twentieth century presented the medical profession with a host of moral issues, but the AMA code has little to say about them. Most of these ethical problems have been raised by developments in science, and the AMA, which speaks primarily for practitioners, has concerned itself more with professional etiquette and the economic welfare of the profession than morality. The maldistribution of medical care, since it affects largely those who can afford neither preventive nor curative medical care, has little bearing on the relationship between physicians and private patients. The first section of the most recent 'Principles of Medical Ethics' asserts that the "principal objective of the medical profession is to render service to humanity," but those who cannot afford private physicians are relegated to crowded, inadequately staffed municipal and state institutions where the care, with some exceptions, can be characterized as third class at best.

University of Maryland,
College Park, Maryland

BIBLIOGRAPHY

1. Anon.: 1831, 'Character and Abuses of the Medical Profession, Rules and Regulations of the Boston Medical Association', *North American Review* **32**, 367–386.
2. Anon.: 1893, 'Differences Between Doctors', *Harpers*, p. 1086.
3. Anon.: 1883, 'The Medical War', *The Nation* **36**, p. 357.
4. Anon.: n. d., 'Percival, Thomas', *Dictionary of National Biography*, Vol. 15, Oxford University Press, Oxford, p. 829.
5. American Medical Association: 1906, *American Medical Directory* . . . , 1st ed., AMA, Chicago.
6. American Medical Association: 1906, *American Medical Directory* . . . , 4th ed., 1914, AMA, Chicago.
7. Andrews, C. S.: 1901, 'Medical Practice and the Law', *Forum* **31**, 542–551.
8. Bell, W. J.: 1970, 'A Portrait of the Colonial Physician', *Bulletin of the History of Medicine* **44**, 497–517.
9. Bonner, T. N.: 1957, *Medicine in Chicago, 1850–1950: A Chapter in the Social and Scientific Development of a City*, American History Research Center, Madison, Wisconsin.
10. Burrage, W. L.: 1923, *A History of the Massachusetts Medical Society, with Brief Biographies of the Founders and Chief Officers, 1781–1922*, Plimpton Press, Norwood, Massachusetts.

11. Burrow, J. G.: 1963, *AMA: Voice of American Medicine*, Johns Hopkins, Baltimore.
12. Burrow, J. G.: 1977, *Organized Medicine in the Progressive Era: The Move Toward Monopoly*, Johns Hopkins, Baltimore.
13. Cordell, E. F.: 1903, *The Medical Annals of Maryland, 1799–1899*, Medical and Chirurgical Faculty of the State of Maryland, Baltimore.
14. Cowen, D. L.: 1964, *Medicine and Health in New Jersey: A History*, Van Nostrand, Princeton.
15. Derbyshire, R. C.: 1969, *Medical Licensure and Discipline in the United States*, Johns Hopkins, Baltimore.
16. Duffy, J.: 1979, *The Healers: A History of American Medicine*, University of Illinois, Urbana.
17. Duffy, J.: 1974, *A History of Public Health in New York City, 1866–1966*, Russell Sage Foundation, New York.
18. Duffy, J. (ed.): 1958, *The Rudolph Matas History of Medicine in Louisiana*, Vol. 1, Louisiana State University, Baton Rouge.
19. Duffy, J. (ed.): 1962, *The Rudolph Matas History of Medicine in Louisiana*, Vol. 2, Louisiana State University, Baton Rouge.
20. Fishbein, M.: 1947, 1969, *A History of the American Medical Association, 1847 to 1947*, W. B. Saunders, Philadelphia, Kraus Reprints, New York.
21. Foshay, P. M.: 1901, 'The Organization of the Medical Profession', *Forum*, pp. 166–171.
22. Jochims, L.: 1979, 'Medicine in Kansas, 1850–1900', *Emporia State Research Studies* 28.
23. Jones, J. H.: 1981, *Bad Blood: The Tuskegee Syphilis Experiment*, Free Press, New York.
24. Kaufman, M.: 1971, *Homeopathy in America: The Rise and Fall of a Medical Heresy*, Johns Hopkins, Baltimore.
25. Kett, J. F.: 1968, *The Formation of the American Medical Profession: The Role of Institutions, 1780–1860*, Yale, New Haven.
26. Konold, D. E.: 1962, *A History of American Medical Ethics, 1847–1912*, State Historical Society of Wisconsin for the Department of History, University of Wisconsin, Madison.
27. Lamb, D. S., *et al.*: 1909, *History of the Medical Society of the District of Columbia, 1817–1909*, Medical Society of the District of Columbia, Washington.
28. Marti-Ibanez, F.: 1958, *History of American Medicine: A Symposium*, MD Publications, New York.
29. New York Academy of Medicine: 1852, *Minutes*.
30. 'Proceedings of the Minneapolis Session': 1913, *Journal of the American Medical Association* 60, 1962–1966, 1989–2019.
31. Reiser, S. J., Dyck, A. J. and Curran, W. J. (eds.): 1977, *Ethics in Medicine: Historical Perspectives and Contemporary Concerns*, MIT, Cambridge, Massachusetts.
32. Risse, G. B., Numbers, R. L. and Leavitt, J. W. (eds.): 1977, *Medicine Without Doctors,* Science History Publications, New York.
33. Rosen, G.: 1946, *Fees and Fee Bills: Some Economic Aspects of Medical Practice in Nineteenth-Century America*, Supplement to *Bulletin of the History of Medicine*, No. 6, Johns Hopkins, Baltimore.
34. Rosenberg, C. E.: 1974, 'Social Class and Medical Care in Nineteenth-Century

America: The Rise and Fall of the Dispensary', *Journal of the History of Medicine and Allied Sciences* **29**, 32–54.

35. Rothstein, W. G.: 1972, *American Physicians in the Nineteenth Century: From Sects to Science*, Johns Hopkins, Baltimore.
36. San Francisco Medical Society: 1868, *Code of Medical Ethics as Adopted by the San Francisco Medical Society, and Recommended by the American Medical Association*, Edward Bosqui, San Francisco.
37. Shafer, H. B.: 1936, *The American Medical Profession, 1783 to 1850*, Columbia University, New York.
38. Shryock, R. H.: 1967, *Medical Licensing in America, 1650–1965*, Johns Hopkins, Baltimore.
39. Wood, H. C.: 1889, 'The Medical Profession, the Medical Sects and the Law', *New Englander and Yale Review* **51**, 118–134.

[illegible] The [illegible] and [illegible] Discovery [illegible]
[illegible]
[illegible] Medical [illegible]
[illegible] University, New York.
[illegible]
Wood [illegible]
[illegible]

SECTION II

MODELS OF THE PATIENT-PHYSICIAN RELATIONSHIP

K. DANNER CLOUSER

VEATCH, MAY, AND MODELS: A CRITICAL REVIEW AND A NEW VIEW*

This is a critical review of two articles which, apparently, have become important and well-known in the literature of the physician-patient relationship. They are: Robert Veatch's 'Models for Ethical Medicine in a Revolutionary Age' [5] and William F. May's 'Code, Covenant, Contract or Philanthropy' [4]. It is out of character for me to devote an article to criticising others. It seems therefore appropriate to mention that I was enlisted in this task by the editor of this volume because of the role these two papers play in the unfolding discussion that has become this volume's theme. I do it in the interests of continuing and advancing this discussion.

These two papers raise a great flurry of issues, any or all of which could be pursued. May's paper in particular – with its rich fund of historical, literary, and theological allusions – could lead a commentator in a multitude of directions. It is therefore important to say that this critique will focus rather single-mindedly on the ethical aspects of the articles. That is, the critique will be concerned with what these models have to do with ethics – what they assume, what they imply, how they deal with and relate to ethics.

Stressing that this critic will focus on ethics may strike many as being superfluous. I suspect the two authors themselves would be so struck, replying something like "But of course these models have to do with ethics; the total issue is one having to do with ethics." Herein lies a crucial point for the understanding of the critique that follows.

'Ethics' is used to denote widely differing domains. For some it seems to include nearly everything – goals, ideals, goods, aesthetics, philosophies of life. For others it is very narrowly circumscribed. (One cannot help thinking that this issue lies at the basis of most disagreements in, at least, applied ethics, though it is seldom addressed head-on.) A lot is at stake in this matter. How we explicate such a concept makes a lot of difference as to what we can do with it. It can be so all inclusive it becomes meaningless; it can be so restricted that it becomes useless. Considerable work must be done to explicate a concept which has had a long and varied use with accumulating connotations. Each step of such conceptual trimming requires an argument. The reader will no doubt be relieved to know that the explication of 'ethics' will

Earl E. Shelp (ed.), The Clinical Encounter, 89–103.

not be argued out as such in this critique, though I believe that our differing views of ethics lie at the base of our differences on the issues dealt with here. These differences should become clear as we proceed.

AN OVERVIEW

Veatch's account of physician-patient relationships is logically and chronologically prior to May's. That is, it is as if May picks up where Veatch leaves off, attempting to advance, widen, and deepen the morally ideal type of relationship. It unfolds something like the following. Veatch briefly describes and subsequently rejects three 'models' of the physician-patient relationship. They are the engineering model, the priestly model, and the collegial model. These are, in a word: 'all facts, no values' (engineering), 'paternalistic' (priestly), and 'buddy-buddy' (collegial). The one, in the end, which Veatch approves of is the contractual model, which embodies the notion of 'contract or covenant'.

May picks up the theme at that point. Distinguishing two types of obligation found in the Hippocratic Oath, he elaborates on the meaning and connotations of each. One is the set of obligations that a physician takes on with respect to his patients, and the other is that which he owes his medical mentors (and their children) for having taught the physician his craft. The first mentioned set (the code) is characterized by gratuitous, perhaps condescending, service rendered to patients (philanthropy). The second set of obligations (the covenant) is characterized as the recognition of a debt by virtue of favors received. It is this second (the covenant) to which May turns most of his attention. The thrust of his unfolding discussion is to show the richness of connotation of covenant and how much more appropriate it is for the physician-patient relationship than either code or contract.

In a sense, then, May is picking up where Veatch left off. Veatch ends by recommending the contractual model; May finds the contractual model inadequate, and recommends the covenantal relationship. Very likely Veatch and May are not in substantial disagreement, if any at all. Veatch did not mean a legalistic contract; he even suggests that he means it more in the sense of "the traditional religious or marriage 'contract' or 'covenant' " ([5], p. 7). However, May explores the concept of covenant much more extensively, resulting in a greater distinction between contract and covenant, whereupon he then opts for the covenantal relationship as the most appropriate for the physician-patient relationship, by virtue of historical, sociological, theological, and – probably – ethical considerations.

VEATCH'S MODELS

It will be best to say at the outset that though these models are colorful, I find them too erratic, whimsical, and uneven to be helpful.

The Engineering Model

The Engineering Model is the one in which the physician sees himself as a 'pure' scientist dealing only with facts, apart from all considerations of value. This presumably must be rejected because the scientist "just cannot logically be value free" ([5], p. 5). That of course is an arguable point. But even assuming the truth of the point, exactly what is its moral relevance? The kinds of 'values' which affect the scientist's conclusions (choice of research design, perceptions, level of significance) are not necessarily *moral* values. They might be aesthetic or perhaps non-empirical commitments of one sort or another. But what then is the 'engineer's' moral failing? Is it self-deception, that is, failing to see that his facts are value-laden? But is self-deception a *moral* failing? If values (of whatever sort) ineluctably taint scientific findings, then what is one to do? Very likely all one can do is to become as aware as possible of those values. But tracing them through medical theories, biochemical theories, microbiological theories and so on to the very foundations of science would be an enormous intellectual task. In fact it may be conceptually impossible, since that task itself would (*ex hypothesis*) be value-laden. Surely Veatch cannot be requiring such measures.

There must be a practical, down-to-earth edge to this issue. Perhaps Veatch is simply emphasizing that values enter into clinical decision-making and that the physician should bring these values to the attention of the patient. This would then be in accord with the good old basic, highly defensible, moral rule, 'Do not deceive'. In that case the Engineering Model would be seen as primarily involving deception, and its moral inadequacies would be obvious.

But apparently that is not what Veatch means to say. For he adds

> ... even if the physician logically could eliminate all ethical and other value considerations from his decision-making – it would be morally outrageous of him to do so. It would make him an engineer, a plumber making repairs ... ([5], p. 5).

It is difficult to see what is morally wrong with this, let alone 'morally outrageous'. Of course if the physician acts with respect to value assumptions without informing or consulting the patient, then the physician (and also, incidentally, the plumber) is morally wrong for either having deceived the patient or for having limited his freedom. But if all value considerations could 'logically be eliminated', where is the moral failing?

In short, there is nothing in this rather abstruse 'engineering model' which could not more clearly and more adequately be expressed in those time honored and rationally based general moral rules "Do not deceive" and "Do not deprive of freedom" [2]. There may be a specifically professional duty (as I shall discuss later) beyond abstaining from deceit and from taking a decision out of the patient's hands, but Veatch does not tell us what it is.

The Priestly Model

The sense of this model is one of paternalism, wherein the physician takes onto himself the decision-making that is properly the patient's. Veatch sees this as 'the opposite extreme' of the engineering model. Presumably the engineering type completely abstained from making value decisions, and the priestly type makes them all. But we still are not sure whether the engineer did not make value judgments, or only thought he did not. If he did not make them, it is not clear that he has been immoral. If he did make them, but unwittingly, then he at least is cognitively deficient, and also perhaps his actions are immoral because he has taken over the decision-making that properly belongs to the patient. But now *that* sounds just like the priestly type, rather than its 'opposite extreme'. Is there in fact a difference, or is it only a difference in style?

There is a further ambiguity as to the real difference between the engineering and the priestly physician. It is not clear whether the priestly one makes value decisions because he believes he has expertise in values, or because he thinks they are factual and not value decisions. If it is the latter, then he is the same as the engineer. Is the 'generalization of expertise' unwitting or not?

However, the chief puzzler in the priestly model is Veatch's claim that the admonition 'Benefit and do no harm' summarizes the priestly tradition and represents the tradition of paternalism. The puzzle may revolve around the ambiguity of 'harm'. I would think that a person is harmed when – among other things – he is deceived, deprived of freedom, and deprived of opportunity. If value decisions that are properly his are made by someone else, I would think he, as a person, is harmed. (Though of course, like any moral infringement, there might be circumstances that would justify it.) Therefore, 'Benefit and do no harm' should *exclude* paternalism, since paternalism would necessarily involve doing harm (or, at least, the belief that one is doing harm), in the sense of 'harm' I have suggested [3]. 'Benefit and do no harm' seems like a sound moral principle, though perhaps too brief to avoid some ambiguity. Rather than incriminating it, Veatch should have focussed more specifically on 'Do not take decision-making away from the patient', for it is that which seems to be the essence of the priestly tradition which he wants to reject.

The Collegial Model

This model is characterized by patient and physician seeing themselves as colleagues, pursuing the common goal of the patient's health. Unfortunately, for this to work we must assume that there would in fact be such mutual loyalty and goals. However, according to Veatch, there is no basis in reality for that assumption, and therefore by virtue of its utopian assumption, he rejects the collegial model.

The Contractual Model

This model's essence is as the name implies, though it is not to be "loaded with legalistic implications." Only in this model, says Veatch, "can there be a true sharing of ethical authority and responsibility" ([5], p. 7). The idea seems to be that patient and doctor would talk out their 'basic value frameworks' in advance, and, if each one is acceptable to the other, they would 'contract' to honor these values on the other's behalf.

On the surface of it, one might wonder what will make this model any more secure than the collegial model. If breakdown of trust and confidence is inevitable on the collegial model, if mutual loyalty and goals are a baseless assumption on the collegial model, why aren't they similarly grounds for rejection of the contractual model? What makes human nature any different on one model than on the other? If trust and confidence are not justified in the collegial model because there is no reason to assume that the participants are truly committed to common goals, why wouldn't the contractual model be rejected for the same reason? It is hard enough to see how having a common goal is necessary, let alone sufficient, for trust and confidence, without having also to see how talking over 'value frameworks' in advance will guarantee a *commitment* to common goals. In short, it seems sheer whimsy to believe that there will be adherence to moral virtues in one case and not in the other. The real working difference between the collegial model and the contractual seems to be the explicitness with which each party makes known to the other his values and goals. That of course is an important point, but it is not the one on which these models have been accepted or rejected – at least not in any straightforward way. Rather they seem to have been accepted or rejected on whether or not the model itself somehow *insures* trust, confidence, moral behavior, and fulfillment of obligations. Yet, this is the very case that has not been made. No reason whatsoever is given as to why, if humans would be immoral in one case, they would not be in another. We can hardly be against an explicit and mutual understanding of values and goals (as in the contractual model) but it certainly is not a sufficient condition for moral behavior and it very likely is not even a necessary condition.

GENERAL REFLECTIONS ON MODELS OF PHYSICIAN-PATIENT RELATIONSHIPS

The central question concerning these models is: why bother? As in so much writing in medical ethics, matters are complicated and muddled beyond any redeeming value. The models are whimsical gestalts which obscure the crucial moral points that could have been made with clarity and crispness.

Given the implicit principles of construction we might have invented many more models: The 'bus driver' model (where the patient knows roughly where the doctor-driver is taking him, he gets on willingly, and he watches the passing scene until he reaches his destination or until something about the passing scene leads him to get off or transfer). The 'pin-ball machine' model (where the patient loudly expresses his emotions and goals and tries everything short of 'tilt' to influence the doctor-machine toward his ends). The 'back-seat driver' model (where there is general agreement about the destination and means of getting there but every inch of the way the patient is telling the doctor-driver how to handle every set of circumstances along the way). There might even be a plumber model – and this would be the ideal – wherein the plumber discusses with his client all the possible value-laden trade-offs within the constraints of the general building code: heat saving devices vs. appearances; less piping vs. slight structural changes; more expense vs. some living style inconveniences. (Why should we ever criticise a doctor for being 'nothing but a plumber'?)

The inventing of models could continue not only *ad absurdum*, but *ad infinitum*. The point is: what is the point of models?

One must assume from the title and subtitle that they have to do with ethics. Yet it is not really clear in any of the models precisely what the moral failing is. Moral points float about here and there but are lost in the miscellaneous details of the model. For example, we saw earlier that in some perfectly good ways of interpreting the engineering and the priestly models they are different only in morally irrelevant ways. At most it is a stylistic difference. That is, there are simply different reasons (and perhaps, causes) for their depriving the patient of his rightful decision-making. And to put the emphasis on the reasons rather than on the actions is to obscure the moral point. The moral point is that patients should not be deprived of freedom or opportunity, which is what happens when decisions affecting them are taken out of their hands.

That point emerges so much more clearly when abstracted from all the rest of the model's conceptual filigree. As they stand, it is not easy to say what is morally wrong with three of the models and what is morally right about the fourth. Mostly they seem to involve mistaken *beliefs*. (The engineer believes there are no values; the priestly believes he is an expert on values, and the collegial type believes he knows the patient's goals and values.) It would seem that empirical argument would be the appropriate response in order to remedy mistaken beliefs, not veiled claims about moral shortcomings.

Consider the engineering model. So much could be inferred from it. If he does not acknowledge that values are intrinsically involved with and determinative of his facts, then he might simply be wrong about a matter of fact. If he knows it, but chooses to ignore it, then he is morally wrong because he is deceiving the patient about something relevant to that patient. But if he is straightforwardly factual with the patient, including such facts as where and how values have determined or influenced conclusions, then it is not clear that he is acting immorally. Indeed, many patients would prefer a physician exactly like that. The point is that the model is a very mixed bag. As such it is not clarifying, not helpful, and not clearly immoral. And I think that is true of all the models.

What would really be helpful is to be told what would be immoral to do to a patient. It is wrong to deceive a patient. It is wrong to deprive a patient of freedom or opportunity (as is done when the physician does not tell the patient of choices or options). These can happen on any of the models; it is mistaken to believe that the 'right' model could prevent it. With this kind of focus, we could let as many models develop as will – as long as no moral rules are broken with respect to the patient. Some patients will prefer the priestly type, and knowingly turn over all decision-making to him. In that case, there is no apparent immorality. The physician-patient relationship would be better served if, instead of delineating models with all their complicated and ambiguous interrelationships, presuppositions, and beliefs, we simply listed what we morally ought not to do – such as deceive, cheat, break confidences and promises or deprive a patient of opportunity or freedom. It is extremely limiting (and maybe immoral) to prescribe the infinitely various details of a physician-patient relationship as though that in itself were a moral matter. Why not let many styles flourish? Let patients and physicians establish the kinds of relationships which suit them. Let them find each other and develop together. The constraints, then, would be the constraints that proscribe certain actions and behavior in the society at large, such as deceiving, breaking promises, depriving of self-determination, and so on.

What is the relation between these models and morality? As we have seen, it is by no means clear-cut. It is highly circuitous at best. However, the *goal* of this delineation of models is relatively clear, by virtue of the article's subtitle: 'What physician-patient roles foster the most ethical relationship?' What is being sought is a format that would most likely maximize morality, a form of relationship that would motivate moral behavior. This assumes a particular view of ethics which is worth exploring. However, since May's arguments boil down (in ethical substance) to this same point, we will look more generally at his article before focussing on it. I see this point as the key link between Veatch and May.

MAY'S CODE AND COVENANT

The reader is referred to the two appropriate paragraphs in 'An Overview' (p. 90) for a quick review of the context and connecting thread of May's in-depth examination of codes and covenants.

May sees codes as very particularized guides to human behavior. That is, they are not universal, binding on all, but molded to meet the needs and fancies of myriad sub-groups of humankind. A code dictates style as well as substance of behavior; demeanor as well as deeds; it bespeaks commitments, techniques, aesthetics, and world views. These codes are fashioned by and for each particular group as an expression of that group's 'philosophy' of those mentioned items.

May finds the code of the medical profession – through its various expressions down through the centuries – to be typical of the codes just described. As the guiding ideal for dealing with patients, the medical code "has not had altogether favorable consequences for the moral health of the profession" ([4], p. 29). The crippling aspect is not what one would expect. Rather, it is the ideal of philanthropy, which, after all, is a very unusual item to be found in these usually very self-serving, self-aggrandizing codes. Nevertheless,

This ideal of service, in my judgment, succumbs to what might be called the conceit of philanthropy when it is assumed that the professional's commitment to his fellowman is a gratuitous, rather than a responsive or reciprocal, act ([4], p. 31).

This is a good insight, which May clearly and convincingly elaborates. There is, on the part of physicians, considerable condescension toward patients; a condescension perpetuated by the code which suggests that physicians owe patients absolutely nothing, and that it is only out of the physician's self-generating goodness that the public is so graciously served. Such an attitude is not justified by the facts, and it does not make for a morally healthy relationship between doctor and patient.

The severe misdirection of the code would incline May toward a contractual relationship as preferable. Among other things, the contract would at least involve informed consent, encourage full respect for the dignity of the patient, acknowledge explicitly the 'symmetry and mutuality' of the relationship, and perhaps provide for legal enforcement of its terms.

Nevertheless, the contractual relationship does not capture the spirit, attitude, and commitment that May would find most ideal. Contracts engender a *quid pro quo* mentality; they lead to minimalism – a doing of no more than absolutely necessary. Focus ends up on the terms of the contract and not on the well-being of the patient. The very nature of health and illness is such that all their related contingencies and surprises could not possibly be exhaustively detailed in a contract. As May so nicely puts it, a contractual relationship

produces a professional too grudging, too calculating, too lacking in spontaneity, too quickly exhausted to go the second mile with his patients along the road of their distress ([4], p. 35).

At this point May turns to the notion of covenant as the most adequate model of the physician-patient relationship. The heart of it is that it is a commitment made (by physicians) in response to gifts and services received (from the public). This is not a tit-for-tat detailing of gifts and responses, but an acknowledgement of overall, undetailable gifts received, so basic and so immense that ledger-keeping of reciprocating responses seems inappropriate. Thus neither gratuitous nor contractual mentalities are apt to surface, though, as May is aware, they *could* surface. That is, if the professional response eventually and obviously outdistanced the 'original' gift, event, or circumstances which provoked the covenant, from then on the reciprocating 'responses' could be seen as gratuitous. Furthermore, if one were led to keep score in this fashion, it would begin to resemble a contract of sorts. One way to avoid both of these eventualities would be to ground the covenant in the transcendent. If the gift is boundless and ongoing, then there is no possibility that response can ever make complete repayment. If you are forever beholding to someone, condescension toward him can never be appropriate.

May has many more interesting insights into the covenantal relationship – its provocations, strengths, and limitations – but for our purposes of focussing on the ethics of the physician-patient relationship, we have drawn out all we need.

COMMENTARY ON CODES AND COVENANTS

It is important to raise the question in a very simplistic way: What is it that May is seeking? It certainly is not mainly a search for what in fact is the relationship between physician and patient, though he spends considerable time on that matter. And he certainly is not seeking a list of right and wrong actions which should or should not transpire between physician and patient, though he deals with the like from time to time. He studies the history, context, and connotations of codes, covenants and contracts. To what end? He rejects this or that aspect of codes, covenants, and contracts, and finds others to be acceptable, even ideal. What are the criteria by which he judges? Criteria and ends are clearly at work. What are they? And what is their relationship to ethics, which is, after all, our main interest?

As we found with Veatch, May is searching for a form, format, or structure of relationship which would create, foster, insure, or motivate ethical behavior on the part of the physician. (Very little if any attention is paid to the patient's behavior toward the physician, though very likely the contractual relationship could embrace that aspect.) Of what is this relationship made up? Beliefs? Attitudes? Imagery? Behaviors? May seems to be consciously building a set of beliefs which will create attitudes which will in turn inspire certain kinds of action. Codes are too ego-expressive and protective, too idiosyncratic, and, in medicine, they bespeak a condescending philanthropy. Contractual relationships lead to a tit-for-tatness and a minimalism. May seems in no doubt as to what results he wants, and all he is looking for is the right set of beliefs to insure those results. May not only wants the physician to do the right thing, but he wants him to do it in the right frame of mind, with the right attitude, and for the right reasons. One gets the impression that May would like to create a wonderful myth which would lead physicians to believe that they were eternally in debt to the people they serve. This would then promote the demeanor and behavior which May finds most acceptable. Again, May apparently already knows how a moral physician acts with patients, and what he is seeking is the right concatenation of truths, beliefs, occurrences, etc. which will ensure that the physician not only acts that way, but does so from the right motives and with the right attitude.

As intriguing as I find all of this with respect to historical, literary, theological, and psychological insights, nevertheless, with respect to ethics and the physician-patient relationship, I find it bewildering and unhelpful. Models seem to be entities or structures which come between patient and physician, fabricated intermediaries which clog the direct relationship between patient and physician (not unlike the traditional metaphysical/ontological problems with the concept of 'relations').

However much the article purports to be about ethics, I think it really is not. I think it really is about motivation and philosophy of life. The model – or relationship – does not tell us what actions are moral nor is it an example of morality but rather its role is to mobilize the physician's attitudes, beliefs, perspectives, and emotions so as to produce moral actions. But 'proper' motivation is neither necessary nor sufficient for moral actions to take place. 'Models' thus are more in the category of sermons, commencement addresses, and other exhortations. Hypnosis, brainwashing, and purity pills might also lead to moral actions. Behind the formulation of models for physician-patient relationship, leading to their acceptance or rejection, has been the implicit consideration: Will it force (or lead, or incline, or motivate) the physician to be moral?

But why such concern for motivation? Morality certainly requires that certain actions be done (or more often, not done) but does not require that they be done from this or that motive, or with this or that attitude, or for this or that reason. It seems an inappropriate criticism of a moral theory to say, "But that would never make anyone be moral." The obsession with motivation I think is misleading and dangerous, unless its distinction from ethics is kept clear. This point needs some elaboration, since it gets at the core of the issue of models of physician-patient relationships.

One guesses that the focus on the physician-patient relationship develops something like this: the relationship is so complicated, there are so many variables, so many different contexts and situations, that we could never spell out explicitly all that a physician should or should not do. Therefore, if we could change his inner self – his attitudes, dispositions, and beliefs – so as to motivate him to be moral, then the myriad individual situations would, by and large, be handled morally. The difficulty with this is that the focus ends up on the inner self – the agent's philosophy of life – more than on what actions are morally acceptable. The criteria for moral-making characteristics of an action shifts to the heart from which it flows rather than to objective moral criteria of the action itself. At that point one is in effect suggesting that any action, so long as it proceeds from the proper motivation, is moral. The motivation might be of the purest sort, say, a realization of profound indebtedness, combined with love and concern. But the like has led not infrequently to an act of unjustified paternalism or deception, or to a broken promise. And those actions are immoral by virtue of criteria external to the motivation which inspired them.

That one's actions are in accord with his own philosophy of life does not make them moral. A philosophy of life concerns what goods that person acknowledges, but bringing about those goods for himself or for others might involve downright immoral actions. A philosophy of life is extremely important, and I think that that is primarily what May is helping us with. It is good, for example, to recognize one's indebtedness to others, to give service unbegrudgingly, to find meaning and commitment in events and encounters, and so on. Some philosophies of life might well be such that they help and encourage us to be moral. The important thing to realize is that what is moral is logically independent of and judged by other criteria than our philosophy of life.

Why not allow these external criteria of moral actions determine what actions in a doctor-patient relationship are moral or immoral? What is the point of creating a new entity – 'the relationship' – which itself, presumably, can be either moral or immoral. Let any kind of relationship flourish, as long as no immorality is done. Interpersonal styles – including mannerisms, dress, tone of voice, eye contact, sense of humor – should not become susceptible to the straightjacket of 'models' which may be embodying 'philosophies of life' in the name of morality. Rather let relationships develop as creatively and freely as they will, being limited only by the immoralities to be avoided.

MORALITY AND PROFESSIONAL CODES: THEIR CONNECTION

It seems only fair and appropriate that I at least sketch the ethical perspective from which I see these issues. Space considerations rule out supporting arguments, but they can be found elsewhere [1, 2]. What follows is grossly oversimplified, but it does give a kind of 'floor plan' to suggest the juxtaposition of some key ideas.

My biggest worry is that ethics has come to mean almost anything. It seems to be an infinitely malleable concept, stretched to fit everyone's whims, goals, and favorite goods. As such it becomes almost meaningless, and discussion involving it becomes almost pointless. Fudge factors prevail. There is in the center of all this, I believe, a hardcore morality, a basic morality which all rational persons would, in a sense, support. Their very rationality would require them "to publicly advocate" this morality ([2], esp. pp. 86–101) lest, as Hobbes would have it, life be poor, nasty, brutish, and short. This basic morality would comprise rules whose central theme is 'Do not cause evil'. The fact that there is a small number of items all rational persons would agree on as evil makes this possible (unlike goods on which we would get very little agreement). The moral rules proscribe the doing of these evils to each other; they are admonitions that would get the public backing of all rational persons. Following these rules requires no effort; it requires only that one does not cause these evils (unlike promoting goods which requires time, effort, risk, and sacrifice). And one can follow these rules universally and impartially, that is, toward everyone, equally, all the time (unlike trying to follow the admonition 'Promote good').

These very basic, 'hardcore', moral rules of course constitute a minimalist morality. But it is a solid beginning, for at least it can have universal and rational support. The sense of all the moral rules is one of proscribing: for example, 'Do not deceive', 'Do not break promises', 'Do not cheat', 'Do not deprive of freedom or opportunity'. However, one of the rules would be "Do your duty" ([2], esp. pp. 121–125). This rule would be publicly advocated by all rational persons because we all come to count on people – firemen, policemen, pilots, waiters, mothers – to do their jobs. If they fail, evil results. So it is to everyone's best interest to urge that prescribed jobs be done, because the rest of us are counting on it.

It is at this point that 'professional codes' enter the scene. Codes, in effect, are delineations of this basic rule 'Do your duty'. Codes spell out the duty of those involved, the ways in which the rest of society can count on them to make efforts, go out of their way, run risks, and make sacrifices. This is not simply the minimalist 'avoid causing evil'; this is more positive and praiseworthy (though it is not the only way to transcend minimal morality).

How these duties get formulated is another matter. Whether it is a self-imposed duty, or one prescribed by the town council, or one that has a long historical tradition, it is crucial that we realize that there are boundaries and limitations – unlike some of the professional codes described by May. A professional code cannot (morally) be simply the self-expression of some group toward any old 'goods' they see fit to pursue. It must be in accord with all the other basic moral rules. That is, the group may promote any goods they want to commit themselves to, as long as they are not breaking any of the other moral rules. Thus, if the physicians' code requires that the physicians do whatever is necessary for the health of the patient, the doctor would still not (morally) be justified in deceiving or in taking away the patient's decision-making role.

It is from this sketched perspective that I see the physician-patient relationship. Moral considerations do not specify the details of that relationship nor the frame of heart and mind the physician must have while doing his job. But it would be important that he not break any of the moral rules with respect to his patients. That of course is no different from our expectations of any person; it is just that, by virtue of the intimate relationship, the physician has so many more opportunities to break moral rules.

What carries the physician beyond the 'mere' doing of no evil is his professional code. Therein is his pledge or his commitment for positive tasks – such as alleviating pain, saving life, comforting the distressed, etc. These then become his duty, people come to depend on these services to be performed, and the physician is morally blameworthy if he defaults on their performance. It is like breaking a promise.

IN CONCLUSION

Constructing models of a doctor-patient relationship seems to be mutiplying entities beyond necessity. It gives us a complicated construct which obscures rather than clarifies the relevant moral issues. Models divert our attention from the morality of deeds and duties to the morality of the model itself. Do we need different models for our relation to our accountant, our barber, our grocer, our lawyer, our architect? Model-talk blurs the fact that physicians have the same moral obligation toward their patients that all humans have toward each other. One can get the impression that if the model is 'in place' in a relationship, all moral obligations are being met, as though programmed by insertion of the model. Models obscure the crucial differences between moral issues and style/manner/personality issues.

Furthermore, the attempt to build incentive and motivation into the model cannot work. Reason will not compel one to be moral but only to advocate morality. It is hard enough to be moral without trying to be moral from the 'right' reasons and the 'right' attitudes. Far better it is to see clearly what the morally right actions are. Let the exhortations to be moral be expressed in sermons, commencement addresses, and the preambles of professional codes.

By seeing the physician-patient relationship as having the same basic moral obligations as any other human relationship, we can then more clearly see the role of professional codes as promises to go above and beyond – *but not in conflict with* – these basic moral rules. And those expressed duties are what we as patients should be particularly alerted to, since those are what makes this professional different from any other. But we have and can have no more assurance that he will live up to those pledges than we have that anyone will live up to their moral obligations. And neither codes, covenants, nor models can make it otherwise.

The Pennsylvania State University College of Medicine,
Hershey, Pennsylvania

NOTE

* This work was supported in part by Biomedical Research Support Grant (NIH) 2 S07 RR05680–13.

BIBLIOGRAPHY

1. Clouser, K. D.: 1978, 'Bioethics', in W. T. Reich (ed.), *Encyclopedia of Bioethics*, Vol. I, Macmillan and Free Press, New York, pp. 115–127.
2. Gert, B.: 1973, *The Moral Rules: A New Rational Foundation for Morality*, Harper Torchbook, Harper and Row, New York.
3. Gert, B. and Culver, C.: 1976, 'Paternalistic Behavior', *Philosophy and Public Affairs* 6 (Fall), 45–57.
4. May, W. F.: 1975, 'Code, Covenant, Contract, or Philanthropy', *Hastings Center Report* 5 (December), 29–38.
5. Veatch, R. M.: 1972, 'Models for Ethical Medicine in a Revolutionary Age', *Hastings Center Report* 2 (June), 5–7.

ROBERT M. VEATCH

THE CASE FOR CONTRACT IN MEDICAL ETHICS

The proposal of four models for the ethical relationship between a health professional and lay person was a very preliminary effort in the development of a more full theory of medical ethics. Thus I am quite sympathetic to many of Danner Clouser's observations about the original formulation of the models.

My defense of what I have called a contract theory of medical ethics had rather accidental origins. The original scheme of the four models was developed as an aside in a dissertation exploring the logic and medical implications of the idea of a value-free science, later published as [3]. At the time I was particularly interested in the rejection of what I called the engineering model, the idea that health care professionals could be counted upon simply to provide the facts so that free, autonomous individuals could supply the values and make, for themselves, clinical decisions. Although now in this era of more enlightened understanding of the relation between facts and values this point may seem obvious to all, in the darker ages of medical ethics of the late 1960s, this was a point that not only needed articulation, it was often rejected out of hand by many in the mainstream of orthodox science.

The short summary of this more extensive discussion that appeared in *The Hastings Center Report* [2] came from an obscure talk I originally prepared for a National House Staff Conference. Had I known that a decade later it would be subject to the scrutiny of a Danner Clouser in one of his most aggressive moments, I surely would have developed the analysis in greater detail.

I am in complete agreement, for example, with his suggestion that the real meat of medical ethics is in the normative analysis, that "the physician-patient relationship would be better served if . . . we . . . listed what we ought not to do – such as deceive, cheat, break confidences and promises or deprive a patient of opportunity of freedom" ([1], pp. 95–96). In fact that is what I and virtually everybody else working in medical ethics over the past decade have been doing. My *A Theory of Medical Ethics* [4] goes on *ad nauseam* doing just that. I disagree, however, with Clouser's suggestion that one can skip over the metaethical preliminaries and jump to normative ethics

Earl E. Shelp (ed.), The Clinical Encounter, 105–112.

straightaway. Unless one deals with the basics of metaethics – of the meaning and justification of moral norms and the role of ethical principles in various moral choices such as professional practices – the normative ethics is likely to be muddled. A full theory of medical ethics requires attention to the basic questions of the place of ethical values and obligation in decision-making. That is why major attention is devoted to such matters in my later work.

In the early days of this generation of medical ethics, I thought it appropriate to begin with a short, schematic typology expressed in terms of models of medical ethical relations between health professionals and lay people. That initial effort, I suggest, was an essential step if we were later to avoid such foolishness as the belief that minimalist ethics simply involves the maxim 'Do not cause evil', and the bizarre notion that a professional code "carries the physician beyond the 'mere' doing of no evil."

THE SYSTEM BEHIND THE MODELS

Clouser suggests in several places that the models are 'erratic', even 'whimsical'. He cannot grasp the system behind them. That is understandable since they have, in their widely circulated form, been detached from the theoretical apparatus that generated them. A theoretical system there was, however. The models were generated out of a rather complex theoretical framework involving two variables I found critical for understanding the role of ethical and other values in medical and other technological decision-making requiring interaction between people with specialized expertise and others lacking it. The two critical variables were (a) the extent to which the professionals are conceptualized as being in a position where they must or should make evaluative judgments and (b) the extent to which there is a convergence between the values of the lay and professional actors. The modeling, heavily influenced by neoParsonian sociological theory of the 1960s, was generated out of the intersecting of these two bipolar variables.

The Necessity of Professional Value Judgments

The ideal type methodology that gives rise to models is, of course, a theoretical construct, and, necessarily, an oversimplified one. The engineering and priestly models were meant to symbolize the problems that can occur if one deviates toward the extremes along the continuum of the first variable, the extent to which professionals are conceptualized as being in a position where they must or should make evaluative judgments. The engineering model was meant to describe the view of those who hold that professionals should not make any such judgments, that they should provide 'just the facts'. The priestly model, on the other hand, symbolizes the conceptualization of the professional role as one where professionals ought to make such value judgments (normally because they have expertise in the values at stake, perhaps because they have been socialized into a special, esoteric knowledge about the relevant values or simply because they have the relevant experience).

It is true that the moral relevance of the models was not made as clear as it might have been in the short, early essay. That was a mission that had to be postponed for additional years and many hundreds of additional pages. It was clear, however, even in the early version that the moral point was not limited to the one identified by Clouser: that physicians involve themselves in deception (self-deception or deception of the patients) if they think they can exclude values from their art. In fact, it was unfair to engineers and plumbers to suggest that their work is value-free or that they view it that way. It is, of course, true that the physician purporting value-freedom has deceived and limited freedom. The moral norms 'Do not deceive' and 'Do not deprive of freedom' are relevant, but something more basic is at stake. The physician who attempts to be value-free, who believes that such a style is either possible or desirable, fails to perceive himself as a responsible moral agent. The more fundamental norm is 'Do not abrogate your moral responsibility for the ethical dimensions of the choices being made'. In the context of the 1960s and the movement for social responsibility of medical and other scientifically trained professionals, this was a critical theme. It was the central theme of the trial of physician army captain Howard Levy, on trial for refusal to use his medical skill to pursue the Vietnam war. It was central to the concern of nuclear scientists asking themselves what their responsibility was for the generation of scientific theory that could predictably be used for devastating harm as well as enormous good. The engineering model gives us a standpoint from which to address the question of the social responsibility of the physician.

The risk of emphasizing the social responsibility of the professional scientist or clinician was that in exhorting to responsibility, excessive moral authority would be claimed, hence, the model at the 'opposite extreme', the priestly model. Clouser is right that the primary moral meat of this construction is the warning against paternalism. The Hippocratic ethic that the physician should use his judgment to do what he thinks is for the benefit of the patient is, indeed, the core of this priestly tradition. It comes from a quasi-religious mystery cult whose members were sworn to secrecy and who pledged by the gods to live their lives in 'purity and holiness'.

Clouser's attack on my formulation of the priestly model is not really targeted on the model itself, but on his claim that the principle of benefitting the patient would require the physician to avoid paternalism since "paternalism would necessarily involve doing harm. . . ." Clouser, here, seems to have been reading too much Mill. Enamored with the principle of avoiding evil, he convinces himself that all conceivable moral wrongs – such as violating autonomy – must involve doing harm. By contrast virtually everyone who attacks paternalism these days does so not simply because violating liberty inevitably does more harm than good but rather because the violation of liberty per se is a wrong-making characteristic. The rule to respect autonomy, for one reason or another, had bite independent of the harm generated by a particular paternalistic act. It is that violation of liberty that priestly-types have been wont to do and that the priestly model warns against.

It is this monomaniac fixation on the principle of avoiding harm that gets Clouser in trouble in defending professional codes of ethics. He defends them on the ground that they are pledges to carry the physician beyond avoiding evil to a commitment to positive tasks such as alleviating pain, saving life, comforting the distressed, and so forth. Clouser views these as unimpeachable goods that are works of supererogation going beyond minimal morality. Then there is no problem if one group unilaterally bestows them upon another without the approval of the other group. If, however, there is a lot more to minimal morality than simply avoiding evil – if, for example, minimal morality includes keeping promises, respecting autonomy, telling the truth, avoiding killing, and promoting justice – then codes by one group bestowing what they presume to be goods upon another without the other's approval may end up actually violating minimal morality.

Either the basic ethical requirements are derived from some moral system shared by lay people and professionals or they are established by mutual agreement. The professional's traditional promise to protect the patient from suffering demonstrates neither of these characteristics. A promise to withhold diagnostic information about a terminal illness in order to spare the patient from suffering carries no moral weight at all unless the patient autonomously accepts that commitment. If the patient prefers to reject it (because it violates his autonomy or violates his dignity) the unilateral conclusion by the profession that such beneficence is morally right counts for nothing. It cannot possibly be a supererogatory act going beyond minimal morality. It does not even come up to a minimal standard required by the principle of autonomy. To be sure a full theory of medical ethics will include the development of the principle of autonomy and the derivative rule 'Do not take decision-making away from the patient', but prior to getting to that point one must have some sense of the moral landscape. That is what the model building was doing when the engineering and priestly models were constructed along the continuum of extent to which the professional must and should make value judgments.

The Convergence of Lay and Professional Values

In many ways the analysis along the continuum of the extent to which the professionals must and should make value judgments is pretty well worn territory. The issues now are closer and closer to being settled. The second variable, however, the variable of the extent of the convergence between professional and lay value judgments, is much richer territory for analysis. Once again I was warning against the danger of two extremes. Once it was realized that a golden mean was necessary between the extremes of value-freedom and value-dominance, some interplay was the obvious solution. That interplay, however, could itself take place along a continuum between one view that saw the lay person and professional sharing the same values (the collegial model) and the opposite extreme where there was a radical polarity between the two value sets. Although I did not give that pole a name, I suppose it might be called legalistic contractualism, the kind of contract one would want to make in a business venture with a used car salesman on the presumption that interests were diametrically opposed.

At the time of the original writing I was much more concerned about the collegial model because it seemed that the danger of falsely believing in the convergence of interests was much greater than the danger of overemphasizing the divergence of interests. The danger of excessive legalism and individualism seemed to me to be obvious to all. That led to the warning against the false presumption of mutual loyalty and trust.

Trained in the tradition of liberal political philosophy with its social contract and in the theological tradition of covenant, which I recognized as providing the historical roots for secular contract, I adopted the term *contract* as a counter to the romantic convergence theory of the collegial model, a term borrowed from Talcott Parsons. Naive as it may seem, it never occurred to me to worry about the legalistic and business-like imagery of the term contract. My image of contract was the bonding of human beings together to form a moral community *à la* Locke or Hobbes. It was the

American founding fathers coming together in a spirit of a compact. It was the covenant that binds man with Yahweh or binds the tribes of Israel together. It was the covenant that bound together the members of the churches of the left-wing of the Reformation of the sixteenth century. It was the contract of a marriage contract, with its almost spiritual overtones. Covenant to me was historically and linguistically a type of contract, the type I have always had in mind for medical ethical relations between professionals and lay people.

Thus Clouser is right when he says that William May and I may not be far apart. There is almost nothing in May's essay to which I can take any exception. If there is any difference at all it would be in my concern that May, Masters, Clouser, and all others who are nervous about contract language are so concerned about protecting the patient-physician relationship from legalism, business imagery, and, to use Clouser's expressive term, "tit-for-tatness" that they overly spiritualize and romanticize the relationship making it as spineless as the buddy-buddy relationship of the collegial model. I have consistently made clear that I oppose individualistic and legalistic implications of the contract metaphor. But I am also concerned about the squishy, apolitical, asocial romanticism of the collegial model. Sometimes covenant is used to buy this softness, and I think it is a dangerous purchase. As long as covenant is seen in its historical vigor – as a political tool for binding a people together as a people, as a compact that generates love and trust and loyalty, but also, when the chips are down, as a metaphor compatible with talk about an eye for an eye and justice rolling down like water – then covenant is the kind of relationship I seek as a way of acknowledging both the convergence of interests between professional and lay person as well as the divergence.

I am just as happy talking about contracts in this way. Marriage is a contract that is fundamentally a relationship of trust and loyalty, of bonding between two people. We surely could ask no more of the patient-physician relationship. That is the kind of contract I have in mind. But we are aware that the marriage contract, spiritualized and romanticized as it is, is embedded in a social, political, and even legal structure. If things fall apart, as unfortunately they do from time to time, there is a socially solid hard core set of institutional structures to fall back upon, not just a mystical, ethereal bond. That is to me what covenant has always required; it is contract at its best.

Clouser asks if the contract model runs the same risks as the collegial model, that "if mutual loyalty and goals are a baseless assumption on the collegial model, why aren't they similarly grounds for rejection of the contractual model?" He misses the difference between the two. The danger of the collegial model is the false sense of security that comes from baselessly presuming a total convergence of interests when such a presumption is not warranted. That means a collegial model will include no checks, no controls on the inevitable divergence that will exist between two people as different in culture, education, goals, and value systems as a physician and a patient. The patient will be a helpless victim of the professional's world view when the professional is making decisions, but equally the professional will be a helpless victim of the patient's world view when the patient is making the decisions. The contractual model (call it the covenant model if you prefer) acknowledges the divergences and has built into it mechanisms to control for them. It emphasizes mutual obligations to inform, the right of either professional or lay person to withdraw when tensions emerge, the necessity of placing the relationship in a larger social context of licensing boards, state laws, etc. In the contractual model trust and confidence are more warranted because they are more limited.

Somewhere out of the blue, Clouser (p. 93) introduces the notion of the virtues, claiming that "it seems sheer whimsy to believe that there will be adherence to moral virtues in one case and not in the other." In the discussion of the models and even the discussion of the normative theory in the years since presenting the models I have never said anything about the virtues. Insofar as moral virtues deal with character traits, motivation, attitudinal characteristics, and other matters fundamentally different from the rightness or wrongness of actions, I am fundamentally in agreement with Clouser when he says that these matters are overemphasized. While I would not go as far as Clouser in saying that these matters can clearly be distinguished from 'ethics', I do agree that they are not the core of ethics, at least insofar as professional ethics is concerned. Thus my models may have little if anything to do with moral virtues.

If I were to develop a theory of moral virtues in the context of the models, I think I would have an easy reply to Clouser's charge, however. It seems obvious that the collegial and contract (or covenantal) models require very different relationships between professional and lay person. Different attitudes, motivations, and character traits would presumably be ideal (although Clouser would be right if he were to say that in neither case would any of these matters be directly linked to the fundamental moral question of right action). Different virtues are appropriate for the two models. The collegial model might be partial to the 'steadfastness' and 'condescension' that were the virtues of Percival's Code and the old AMA Code of 1847 or to the 'filial piety' that was the virtue of traditional Chinese medicine. In contrast the contractual model might be more inclined toward the virtues of respect, faithfulness, and (to the extent it is a virtue) integrity. If the virtues are different in the two models and some virtues are more easily adhered to or are more worthy of adhering to than others, then Clouser's problem seems answered.

CONCLUSION

In short, the models of ethical medicine were early, preliminary attempts to deal with some basic questions in a metaethic underlying medical ethics. I never dreamed that they would be seen as functioning in place of a full normative ethical theory. I never even dreamed that they would stand alone as a full metaethical theory. They, however, hint at some fundamental problems in medical ethics, certainly a richer array of problems than Clouser recognized. Moreover they were built on a general systematic theory such that these models and only these models (plus legalistic contractualism) should have been expected to result. Of course, the models by themselves are nothing like a full ethical theory for medicine. They do give insights, however, into some of the major problems for an ethical theory in medicine. They show why medical ethical codes generated exclusively by professionals must fall within the priestly model, why they are philanthropy, to use May's term, and why philanthropy – the bestowing of an ethic unilaterally by one group upon another – is likely to run into difficult times.

Kennedy Institute of Ethics,
Georgetown University, Washington, D. C.

BIBLIOGRAPHY

1. Clouser, K. D.: 1983, 'Veatch, May, and Models: A Critical Review and a New View', in this volume, pp. 89–103.
2. Veatch, R. M.: 1972, 'Models for Ethics in Medicine in a Revolutionary Age', *Hastings Center Report* 2 (June), 5–7.
3. Veatch, R. M.: 1976, *Value-Freedom in Science and Technology*, Scholars Press, Missoula, Montana.
4. Veatch, R. M.: 1981, *A Theory of Medical Ethics*, Basic Books, Inc., New York.

K. DANNER CLOUSER

A REJOINDER

I certainly understand Veatch's dismay over having to deal with a "preliminary effort" he wrote more than ten years ago. My empathy for such was part of my reluctance to criticize it in the first place. I trust it is clear that my critique was of the particular article only and not of Robert Veatch or of any of his other work. And since the particular article in question has apparently become something of a classic, the editor deemed it an appropriate launch pad for further discussion of the doctor-patient relationship. For what it is worth, it might be noted that my criticism does not really use its ten year vantage point. I would have leveled the very same criticisms had I written my reply ten years ago, for they have to do with logical and theoretical matters rather than with the social or historical context.

A rejoinder must be brief. Time cannot be spent on ironing out small disagreements and misunderstandings. I see Veatch's reply falling into two categories: about three-fifths of it discusses his models, and two-fifths of it attacks (what he takes to be) my view of professional ethics. Concerning the models Veatch says a lot about the how, when, and why of their formulation. Though these observations are of some autobiographical interest, there is nothing there that leads me to change any of my criticism of them. I will, however, briefly reply to several points that might be thought to make a difference, and then I will deal briefly with his criticism of (his perception of) my view of professional ethics. Though my view of professional ethics emerged only piecemeal and then mostly in response to William May's article, since Veatch has taken it on, it becomes fair game for discussion. Besides, I do think it would be the most fruitful focus at this point in our brief exchange of ideas.

There are four small points with reference to Veatch's reply about the models which I want to address.

(1) Veatch says (p. 105) "I disagree, however, with Clouser's suggestion that one can skip over the metaethical preliminaries and jump to normative ethics straight away." I do not see that I either said or implied such a thing. Indeed, as one who does not write about medical ethical issues *unless* they involve some matter of ethical theory, I am particularly baffled by Veatch's comment. I do of course think that straightforward, obvious moral rules are a

Earl E. Shelp (ed.), The Clinical Encounter, 113–116.

lot clearer than the models, but then I did not see the models as any kind of moral theory, or even as "metaethical preliminaries".

There may be an important metaethical matter at stake, however. I do believe that before one can have even an inkling of a valid ethical theory, one must examine moral phenomena – the way morality is lived and talked and reasoned about. And I think this happens to center on admonitions which are paradigmatically *moral* admonitions, and that there is no other way to individuate the moral realm, and hence no other way of determining the data from which to begin theorizing. Those philosophers who wittingly or unwittingly formulate their ethical theory first are apt to end up with idiosyncratic theories which are more conceptual castles in the sky than systematic accounts of the moral point of view.

(2) Veatch's criticism of my suggestion that the admonition "benefit and do no harm" (p. 108) might exclude paternalism is easily dealt with simply by referring him to what I actually said (p. 92). There I made it clear – indeed, explicit – that I was using 'harm' as a shorthand expression for such things as deceiving and depriving of freedom and of opportunity. Those are 'harms' or 'evils' by virtue of the fact that all rational persons would avoid them unless they had a reason not to. And one cannot be paternalistic without doing harm in that sense.

(3) Concerning the Collegial Model, Veatch says (p. 110) that I miss the difference between the collegial and the contractual model when I say "if mutual loyalty and goals are a baseless assumption on the collegial model, why aren't they similarly grounds for rejection of the contractual model?" But I wonder in return if Veatch is missing the difference between what he says and what he wants to say. Look at what he says about the collegial model ([2], p. 7).

(a) Patient and physician see themselves as colleagues pursuing the common goal of eliminating the illness and preserving the health of the patient.

(b) The physician is the patient's 'pal'.

(c) The theme of trust and confidence play the most crucial role.

(d) When two individuals or groups are truly committed to common goals, then trust and confidence are justified and the collegial model is appropriate.

(e) There is an equality of dignity and respect and an equality of value contributions.

Those features seem to constitute the collegial model. Now, look at Veatch's criticism, namely, that, in fact, there is no real basis for the assumption of mutual loyalty and goals or of common interest. But who said anything about *assumption*? *How* professionals and clients get their common goals and interests was not mentioned as part of the model itself. 'Mutual loyalty' also is not constitutive of the model itself. It is a personal virtue which may or may not occur with the collegial model; it is external to the model itself. Now, one of the failings of this model is that it needs loyalty in order to work, but there are no grounds for mutual loyalty. And my original question was: if we cannot assume mutual loyalty for this model (in order to make it work), why wouldn't it also be a failing for the contractual/covenantal model (which surely would need to assume mutual loyalty in order to work)?

I think that Veatch really meant the collegial model to be one in which professional and client simply *assume* that they share the same values, interests, and goals – and never ask, discuss, or compare. And then it fails because in fact they are not really the same. (But notice mutual loyalty would still be possible.) However, since he did not say it that way I can't be sure.

(4) Veatch accuses me of "somewhere out of the blue" introducing the notion of moral virtues, where I write "It seems sheer whimsy to believe that there will be adherence to moral virtues in the one case and not in the other" (p. 93). I am guilty of putting the two words 'moral' and 'virtue' side by side, but not of introducing the notion of 'moral virtue'. Its being "out of the blue" should have been his first clue that in this context I was not dealing with the technical sense of moral virtue. I could have said without loss of meaning "... that there will be adherence to morality ...". I had simply wanted to cover all bases for whatever Veatch was regarding as moral. In that context he had been talking about such diverse things as loyalty, trust, confidence and moral behavior. So "moral virtue" seemed a good general expression for whatever moral behavior Veatch thought could or should be insured by the model. I no more meant to introduce the concept of moral virtues thereby than Veatch, when he mentions (out of the blue) "moral meat," meant to introduce the concept of animal rights! (p. 107).

We now turn our attention to Veatch's attack on my comments about professional ethics. The text for this discussion might well be his statement:

> It is this monomaniac fixation on the principle of avoiding harm that gets Clouser in trouble in defending professional codes of ethics (p. 108).

Veatch is right about one thing: I do believe that if one is going to be a maniac, he should be singleminded about it! But from that point on, Veatch so muddles the points and distinctions that I don't even recognize it as a *position*, let alone as mine.

In this context the basic issue between us is this: He believes it is wrong for professionals unilaterally to dream up goods (articulated in their codes and the like) and then unilaterally to force those goods on their clients whether the clients want them or not. Furthermore, Veatch thinks that I am committed to approving of such behavior. But of course I am not.

Rather than repeat what I actually said, I refer the reader to pages 92–93, 100–103 of my critique. The answers are there. However, underscoring of certain points is obviously in order. There is an important distinction between "preventing evils" and "promoting goods," which Veatch has missed. The former is a moral ideal and sometimes justifies breaking a basic moral rule; the latter is not and does not. Professionals (on my view) would spell out their duty (that is, their commitment above and beyond the basic moral rules), in terms of moral ideals (that is, in terms of preventing evils). Notice that that is very different from laying "unimpeachable goods" (p. 108) on the clients. An evil (by definition) is something that all rational persons would want to avoid, unless they had a reason not to. So there is no moral problem with unilateral prevention of evil. Of course, there can be situations in which there might be a conflict between preventing an evil and breaking a basic moral rule. And then one cannot automatically break the moral rule in order to fulfill the moral ideal. In that case, though the preventing of evil is a crucial consideration in justifying the breaking of a moral rule, one must still follow the usual procedure of moral reasoning necessary to justify an infringement of a moral rule (for example, see [1], Ch. 8).

So, in answer to Veatch, there is basically no moral problem with unilateral prevention of evils, and there cannot be unilateral bestowing of goods on another without the other's approval. That would be depriving them of opportunity and freedom, and perhaps of pleasure, if not straightforwardly causing mental or physical pain. In short, it would be breaking some basic moral rules.

To quote from the last paragraph of my critique:

> By seeing the physician-patient relationship as having the same basic moral obligations as any other human relationship, we can then more clearly see the role of professional codes as promises to go above and beyond – *but not in conflict with* – these basic moral rules (p. 103).

The Pennsylvania State University College of Medicine,
Hershey, Pennsylvania

BIBLIOGRAPHY

1. Culver, C. M. and Gert, G.: 1982, *Philosophy in Medicine: Conceptual and Ethical Issues in Medicine and Psychiatry*, Oxford University Press, New York.
2. Veatch, R. M.: 1972, 'Models for Ethical Medicine in a Revolutionary Age', *Hastings Center Report* 2 (June), 5–7.

BARUCH A. BRODY

LEGAL MODELS OF THE PATIENT-PHYSICIAN RELATION

I will in this essay attempt to do three things. First, I will try to sketch two ways of thinking about, two models of, the patient-physician relation. Secondly, I will try to show how these two different models are in effect presupposed by two very different legal treatments of the patient-physician relation, the treatment of that relation in most common-law jurisdictions and the treatment of that relation in Judaic Law. Finally, I will try to show how certain different fundamental values lie behind these models and to suggest that a fully adequate moral model needs to find a way to accommodate these different values.

MODELS OF THE PATIENT-PHYSICIAN RELATION

The first model of the patient-physician relation is what I shall call the commercial contract model. It makes five basic claims about that relation, viz, that neither party in that relation is under an obligation to enter into it, that each party enters into it only on the basis of the terms to which it freely agrees, that the relation legitimately exists only when the terms set in advance by each party are the same, that society's main role in connection with this relation is to protect both parties against non-performance or malperformance and against fraud and/or coercion, and that society does not pay the agreed-upon fees for the services rendered the patient by the physician. Even this brief statement makes it clear why this model is properly called the commercial contract model. According to this model, the patient-physician relation is best viewed as a commercial relation created by normal contractual processes. The aptness of this name will be clearer, however, after we elaborate upon each of these five claims.

The first is that neither party in the physician-patient relation is under an obligation to enter into it. This means, of course, two things. The first is that no physician is under an obligation to provide services to patients in general or to some particular patient. This is true even if there is an emergency situation in which help is needed immediately. A physician who fails to offer his services, especially in an emergency situation, may perhaps be open to criticism on a wide variety of grounds (e.g., lack of compassion), but such a

Earl E. Shelp (ed.), The Clinical Encounter, 117–131.

physician cannot properly be criticized on the grounds that he has failed to fulfill his obligations as a physician. The second is that no potential patient is under an obligation to seek a physician to treat his medical problems. This is true even if the potential patient is in serious need of medical treatment. Again, a potential patient who fails to seek medical help, especially when it is seriously needed, may be open to criticism on a wide variety of grounds (e.g., stupidity), but such a potential patient cannot properly be criticized on the grounds that he has failed to fulfill an obligation he has to seek that help.

The second of these claims is that each party in the physician-patient relation enters into their relation only on the basis of terms to which he has freely agreed. This means, of course, different things for the physician and for the patient. For the physician, it means that he is under no obligation to provide services at some set time, or in some set place, or in some set manner, or for some set fee. He is free to provide services in accordance with those terms which are acceptable to him (providing, of course, he can find someone who wishes these services subject to those conditions). For the patient, it means that he is free to accept only such medical services as are performed in accordance with such terms as to time, location, manner, and price as are acceptable to him (providing, of course, he can find someone who wishes to provide the services subject to those conditions). He is under no obligation to accept the terms offered by some physician.

The third of these claims is that the patient-physician relation legitimately exists only when both parties have agreed to the same terms governing the provision of these medical services. This third claim, which is the heart of the commercial contract model, follows from the first two claims and from the general rules governing commercial relations providing one accepts, as does this model, the idea that the patient-physician relation is essentially a commercial contractual relation. One caveat about this third claim. Even in the commercial case, a relation can in extraordinary circumstances be created unilaterally. If, for example, your property is threatened by destruction and I can save it but you are unavailable to consent, then if I do save it, I can obtain certain payments from you. We may, in such cases, allow the relation to come into effect on the basis of your hypothetical consent, the consent you would have given had you been able to consider the offer of my services. Similarly, in the medical context, we can even on this model say that a physician-patient relation also exists when a physician offers his services in an emergency situation in which the patient's consent cannot be obtained. The legitimacy of the relation is based upon the physician's actual agreement and the patient's hypothetical agreement.

The fourth of these claims is that society's main role in connection with the physician-patient relation is to protect both parties against non-performance or malperformance and against fraud and/or coercion. This conception of the social role in connection with the physician-patient relation is the same as the classical liberal conception of the social role in connection with any commercial transaction. Society must, to begin with, protect people against the use of fraud or force to obtain consent to relation-creating agreements. We need to make sure that the consent of both parties is freely obtained after full disclosure of the relevant terms of the agreement. In recent years, these requirements have been strengthened as we have come to understand more about subtle fraud and subtle coercion, but the basic idea behind this part of the model is very old. Society must also protect both parties against non-performance or malperformance. Malperformance gives rise to a tort suit and non-performance to either a suit for damages or for performance. In theory, these social protections are protections for both parties. In practice while both parties are protected against non-performance, the protection against malperformance is primarily a protection for the patient against the physician; it is hard to imagine how a patient can malperform his role as opposed to non-performing.

The fifth and final claim of this model is that society does not pay the agreed-upon fees for the services rendered by the physician. This important aspect of the model grows out of the model's insistence that the patient-physician relation is created by agreement by those parties to a set of terms, including terms governing payment for the services provided. The patient's agreement can oblige him to pay certain fees; it is unclear how it can create a social obligation to pay those fees, and the commercial model insists that there is no such obligation. All of this is perfectly compatible with the idea that some third party (e.g., an insurance company) can have an obligation to reimburse the patient for the fees paid, an obligation created by some agreement in advance between the patient and the third party. All that is being claimed is that, absent such an agreement, the obligation to pay the fees falls upon the patient.

So much for the basic claims of the first model. We turn now to the second model of the patient-physician relation which I shall call the status model.[1] Its basic claims are that both parties to the patient-physician relation in certain conditions are under an obligation to enter into that relation, that the terms of the relation are set by certain objective factors which are independent of and may conflict with the wishes of one of the parties, that the relation legitimately exists when the above-mentioned conditions hold between patient and physician, that society's main role is to insure that each party enters into this relation when they are obliged to do so and to insure that the relation is governed by the appropriate terms, and that this may entail society's paying for the appropriate fees. The contrast with the commercial model is clear even on the basis of this brief statement of the status model. The contrast will be even clearer, however, after we elaborate upon each of these five claims.

The first is that both parties to the patient-physician relation are under obligations to enter into that relation under certain conditions. This claim means different things when it is applied to the patient and when it is applied to the physician, for the conditions differ as applied to each. For the potential patient, this claim means that the potential patient is under an obligation to seek out a physician's help when he has a medical problem, and the obligation becomes greater as the problem becomes more serious. A potential patient is not entitled to refuse to become a patient if he needs medical help. On different versions of this model, the obligation may be viewed as an obligation to the patient himself, or to society, or to God, or as not being directed to anyone; on all versions, however, the potential patient fails to fulfill one of his obligations when he fails to seek adequate medical care for his medical problems. For the physician, this claim means that he is under an obligation to provide services to patients in general and to particular patients. This is certainly true in emergency situations. A physician who fails to perform in emergencies can properly be criticized on the grounds that he has failed to fulfill his obligations as a physician. On most versions of this model, this obligation also holds in non-emergency situations as well, although the precise nature and extent of this obligation in non-emergency situations varies from one version of this model to another.

The second of the claims of this model is that each party to the patient-physician relation enters into that relation under terms which are at least in part set independently of their wishes. Again, this means different things for the physician and for the patient. For the patient, the most important thing that this means is that there are constraints on the types of treatment and conditions for treatment the patient can request. The most important constraint is that the treatment must be one of those which are judged most appropriate by professional opinion. According to this second claim, even if a potential patient can find a physician willing to provide services of questionable merit, the provision of such services is wrong, is not in accord with the proper terms governing the patient-physician relation. For the physician, the most important thing that this means (in addition to those constraints on forms of treatment) is that there are defined limits on fees and defined requirements concerning availability, etc. Even if there are patients willing to meet the requests of the physician in these areas, the rules governing proper patient-physician relations require that the physician not charge the higher fee, etc.

The third claim is that a legitimate patient-physician relation is one governed by these obligations and these terms. This is the heart of the status model and is what really differentiates it from the commercial model. Even if patient and physician agree to some other rules (say the provision of questionable forms of treatment for higher fees), their agreement cannot turn the resulting relation into a legitimate patient-physician relation. Legitimacy, according to this model, is based upon the content of the relation, not the consent to it. Similarly, there can be, according to this model, a legitimate patient-physician relation even when one party has not consented (and even objects) to its very existence. A physician providing an appropriate treatment to a patient in need of that treatment has created a legitimate relation even without the patient's consent since the content of the relation determines legitimacy. Emergency medical treatment, on the status model, requires no hypothetical consent to justify it.

The fourth claim of this model is that society's main role is to insure that each party enters into this relation when they are obliged to do so and to insure that the relation is governed by the appropriate terms. The emphasis on society's protecting against fraud and/or coercion is appropriate in the commercial contract model because the fraud and/or coercion invalidates the consent which is the foundation of that model of the legitimacy of a particular physician-patient relation. This emphasis would be inappropriate in the status model since it grounds legitimacy in content rather than consent. (Remember that in this model a patient can legitimately be treated without consent.) Since it does, however, it is naturally led to the conclusion that society's main role is to protect legitimacy by insuring that the relevant parties have entered the medical relations they are obliged to enter on the required terms. It should be noted that this includes protecting against non-performance and malperformance.

The final claim of the status model is that society's role may entail its paying for the medical treatment. It is easy to see how the last claim arises out of the earlier claims of the status model. Suppose we have a case in which society has compelled an indigent patient to obtain needed medical treatment and has required a physician to provide that treatment. Who is to bear the cost of the physician's expenses and time? It cannot be the patient because he doesn't have the capacity to pay. We might impose the costs on the physician, but it is unclear why it should be his burden alone and in any case he is likely to pass it on to us through his fees. Realism and a sense of fairness seems to lead to the conclusion that society must sometimes bear the cost.

In short, we have seen in this first section how there are at least two radically different models for the patient-physician relation. In the next section, we will see how they are embodied in different legal systems.

LEGAL SYSTEMS AND THE PATIENT-PHYSICIAN RELATION

In this second section of the paper, I will be arguing that the common-law treatment of the physician-patient relation is captured by the commercial contract model, while the status model captures the treatment of that relation in Judaic Law. But before arguing for those claims, I need to say a little bit more about what it means to claim that some model captures the treatment of the physician-patient relation in some legal system.

There are several things that such a claim does not entail. It does not entail (although it may be the case) that jurists or legal scholars in the system in question have explicitly formulated the whole model or even each of its important claims. It even does not entail (although it may be the case) that all important decisions in that legal system covering the patient-physician relation are in accord with what that model would indicate. All that is entailed by that claim is that most of the leading decisions in the system covering the patient-physician relation are in accord with what that model would indicate.

With that understanding in mind, let us turn to the various claims made by the commercial-contract model concerning the patient-physician relation. In each case, we will try to show either how it is explicit in most of the relevant leading decisions and laws in common-law jurisdictions or at least how most of the decision and laws are in accord with what the commercial-contract model would indicate.

The first of those claims was that neither party in the physician-patient relation is under an obligation to enter into that relation. The common-law treatment of the physician side of this claim is well known; even in emergency situations, the common-law imposes no requirements on the physician to treat. In fact, until recent years, prudence dictated to physicians that they not stop to help patients in emergency situations since they might expose themselves to liability for malpractice while they were under no threat of liability for failing to stop since they have no obligation to stop and help.

This anomaly has led many states to pass 'Good Samaritan' statutes[2] protecting physicians who help in emergency cases from the threat of liability for malpractice and has led states to consider requiring by statute such emergency aid. The common-law treatment of the patient side of this claim is more complicated. Cases arise with less frequency, since who will sue to require a prospective patient who doesn't want to be treated. The most frequently heard cases are those involving patients already being treated who want to discontinue treatment. Even in such cases, e.g., cases of adult Jehovah's Witnesses wishing to discontinue treatment because they do not wish blood transfusions,[3] many courts have held that the patient is under no obligation to continue being treated.[4] And, in fact, even those courts which have required continued treatment against the will of the patient have stressed that aspect of its being continued treatment, although they have been ambiguous about what would happen if the patient were not yet being treated. Thus, Chief Justice Weintraub wrote:

> When the hospital and staff are thus involuntary hosts and their interests are pitted against the belief of the patient, we think it reasonable to resolve the problem by permitting the hospital and its staff to pursue that function according to their professional standards. The solution sides with life, the conservation of which is, we think, a matter of state interest [5].

The second of the claims of the commercial-contract model was that each party in the physician-patient relation enters into that relation only on the basis of terms to which he has freely agreed. On the physician side, this time, there is little common-law treatment of the issue. It is taken for granted, and it is only in recent years that society has begun to speak (very weakly) about fees, places in which he practices, etc. There is more case-law on the patient side, and the traditional principles are clear. The whole doctrine of consent presupposes that the patient must know and freely consent to what is being done to him in order that the physician's act not be a battery. And case law makes clear that the terms of the patient's consent structures the physician-patient relation. In the classic case of *Mohr v. Williams* (in which the doctor operated on a different organ with a different disease than the one for which he had obtained consent), the court stated:

> The last contention of the defendant is that the act complained of did not amount to an assault and battery. This is based upon the theory that, as plaintiff's left ear was in fact diseased, in a condition dangerous and threatening to her health, the operation was necessary, and, having been skillfully performed at a time when plaintiff had requested a like operation on the other ear, the charge of assault and battery cannot be sustained. . . .

> We are unable to reach that conclusion . . . every person has a right to complete immunity of his person from physical interference of others . . . and any unlawful or unauthorized touching of the person of another, except it be in the spirit of pleasantry, constitutes an assault and battery [8].

No doubt, recent decisions have been more protective of the physician. More recent courts have been willing to say:

> The consent – in the absence of proof to the contrary – will be construed as general in nature and the surgeon may extend the operation to remedy any abnormal or diseased condition in the area of the original incision whenever he, in the exercise of his sound professional judgment, determines that correct surgical procedure dictates and requires such an extension of the operation originally contemplated. This rule applies when the patient is at the time incapable of giving consent, and no one with authority to consent for him is immediately available [6].

And this extension has certainly been aided by blanket consent forms used by physicians and hospitals. Still, even these recent tendencies operate within the theses of the commercial-contract doctrine,[5] since (a) clear-cut limitations put on by the patient must still be observed, (b) generalized consent only applies to that area of the body, and (c) the patient's explicit consent for the change must be obtained when possible.

The third of the claims, that the relation legitimately exists only when the terms set in advance by each party are the same, follows, as we saw before, from the first two claims. We turn then to the fourth claim of the commercial-contract model, the claim that society's main role in connection with this relation is to protect both parties against non-performance or malperformance and against fraud and/or coercion. The common-law has clearly assigned this role to the law in policing the physician-patient relation. Physicians, as we all know too well, can be sued for the damage caused by their non-performance and/or malperformance and they in turn can use the law to help them when patients non-perform by refusing to pay their fees (we leave aside for now the question of the economic feasibility of either party suing). Moreover, the law has acted as a vehicle of protection against fraud. Licensing requirements and the Pure Food and Drug Acts[6] certainly come to mind in this connection, and these forms of protection have traditional as well as contemporary guises.

The last of the claims of the commercial-contract model is that society does not pay the agreed-upon fees for the services rendered the patient by the physician. The patient does so, either directly or through some insurance scheme, or the physician treats without fees. This is, of course, the area in which common-law jurisdictions are now moving most rapidly away from the model. In the United States, the government is increasingly the source of payment for medical services; in other common-law countries (e.g., England), it has been the predominant source for some time. These developments are, however, very recent. Until recently, individuals paid for their medical services. And it may well be suggested that much of the uneasiness about the medicine-law relation today is precisely connected with this sudden change in the source of payment. With thesis five under challenge, we can expect what we are observing, increasing challenges to the earlier theses (at least in the case of those physician-patient relations where society is the source of payment). In any case, until these recent developments, thesis five certainly held.

So much for the ways in which the commercial-contract model captures the common-law treatment of the physician-patient relation. We turn now to the ways in which the status model captures the Judaic Law treatment of that relation.

The first of the claims of that model is that both parties to the patient-physician relation in certain conditions are under an obligation to enter into that relation. The obligation of the physician, particularly in emergency situations, is part of a general obligation which Judaic Law imposes upon people to save others in need:

> How do we know that if someone sees another drowning in a river or pursued by a wild beast or being attacked by brigands that we are obliged to save him? We learn it from the biblical verse 'you should not stand by the blood of your neighbor' (*Sanhedrin* 73a).

The ill patient is also threatened by his illness, and the physician, like everyone else who can aid the patient, is obliged to do so. Although this extension of the principle is not explicitly mentioned in the Talmud, Maimonides develops it in his code (*Mishneh Torah, Hilchot Nedarim* 6:8). A similar type of development arises in the case of the patient. Judaic Law prohibits people putting themselves in conditions that are dangerous to their life and/or health. Since refusing treatment is equivalent to placing oneself in such a condition, the clear implication is that patients requiring medical treatment cannot refuse the necessary treatment. And most rabbinic authorities have drawn just this conclusion.[7] I note, *en passant*, that the roots of both of these obligations, the physician to treat and the patient to be treated, presuppose that there is in the condition at least a remote threat to the life of the patient. Medical care involving no such threat (e.g., cosmetic surgery) seems therefore to be obligatory neither for the patient nor for the physician.

The second of the claims of this model is that each party to the patient-physician relation enters into that relation under terms which are at least in part set independently of their wishes. For the physician, there are limitations upon when he can treat, how he can treat, how much he can charge for the treatment, etc. To get a flavor of the sort of limitations, one might look at the following passage from Epstein's *Aruch Hashulchan*:

It is forbidden to practice medicine unless one is an expert who has permission from a court to practice . . . and also one should not practice if another more expert physician is present. . . . And as far as fees are concerned, the Tur writes in the name of Nachmanides that one can be compensated for the time and effort expended but not for the study one has first undergone . . . as God says about doing good deeds, much as I act without a fee, so should you act without a fee (*Yoreh Deah* 336:2–3).

In fact, this last limitation obviously posed serious difficulties, and Nachmanides (*Torat Ha' Adam, Shaar Ha' Sakanah*) already indicated that if the patient has agreed to pay a fee which was higher than compensation for time and effort (leaving aside the question of how those are to be determined), the patient is obliged to fulfill his agreed-upon obligation. This opinion and the various issues it raises has provoked immense discussion; the crucial point for our purposes is that the presupposition of the whole discussion was that the normatively preferable situation was one in which the fee was set by objective factors independent of contractual negotiations between the patient and the physician. For the patient, there are limitations upon the types of treatment which he can accept and refuse. A difficult question, which has been much discussed, is the moral legitimacy of a patient undergoing a risky and/or experimental therapy[8] in cases ranging from those in which his life is threatened and nothing else is available to those in which he is merely seeking to alleviate pain. In this long debate, the question has never been settled by reference to what forms of therapy the patient and physician are willing to try; the legitimacy of a given form of treatment raising such issues is settled on different grounds.

All of what we have said so far presupposes the truth of the third of the claims we have identified in the status model, the claim that a legitimate physician-patient relation is one governed by these obligations and terms, and not one which is merely based upon mutual consent. Judaic Law would object, to use our earlier example, to the provision of questionable and dangerous treatments for higher fees even if both patient and physician accept these terms, in part because the form of treatment undercuts the legitimacy of the relation and in part because the fee structure undercuts its legitimacy.

We turn then to the fourth claim of this model, the claim that society's main role is to insure that each party enters into this relation when they are obliged to do so and to insure that the relation is governed by the appropriate terms. There is less direct material to cite here; all of the cases alluded to above presuppose the truth of this fourth claim. The fact that patient and/or physician do turn to the court to tell them how their relation should be structured and the court responds by doing so presupposes that society (in the form of the court) feels obliged to oversee the terms of the relation between physician and patient.

In many ways, the most interesting question is the extent to which Judaic Law accepts claim five, the claim that society's role may entail its paying for the medical treatment. One thing is eminently clear. The physician has the obligation to treat, even if he cannot get paid by the indigent patient. This follows directly from the analogous principle about our obligation to save others even if we have to incur costs to save them (*Sanhedrin* 73a). Similarly, the community has an obligation to save those who are ill much as it has an obligation to save those who are captured by brigands and who are facing a threat to life (*Yoreh Deah* 252). In fact, this is a paramount obligation. The tricky question, which I have not found explicitly discussed, is whether the physician who treats the indigent patient can turn to the community and demand that it bear the cost.

Leaving aside this last question, we can, I believe, safely conclude that Judaic Law is committed to the main claims of the status model of the patient-physician relation, much as the common law is committed to the main claims of the commercial-contract model. There are, then, more than one legal model of the patient-physician relation. We will, in the final section of this paper, look more critically at these two models, and see what can be said by way of combining their strengths.

VALUES AND MODELS OF THE PATIENT-PHYSICIAN RELATION

In this final part of the essay, I wish to say a little bit more about the values that seem to lie behind these different models, about the strengths and weaknesses of each of these models, and about one common feature of both of these models that may need to be rejected if we are to develop a comprehensive model of the patient-physician relationship.

It is clear that the fundamental value lying behind the commercial contract model is that of the autonomous free choice of both parties in the patient-physician relation. From the point of view of physicians, this model preserves their freedom to choose when and how to practice. From the point of view of the patient, this model preserves their freedom to choose when and how to be treated. To the extent that we find this model attractive, we do so, I believe, because we place a high value on this freedom and autonomy. It is also clear that the fundamental value lying behind the status model is that of responsibility. From the point of view of physicians, this model emphasizes their responsibility to treat patients in an appropriate fashion for a reasonable fee. From the point of view of patients, this model emphasizes their responsibility to seek the best medical treatment required to maintain the highest level of health possible. From the point of view of society, finally, this model emphasizes responsibility to fund the medical treatment required. To the extent that we find this model attractive, we do so, I believe because we place a high value on these responsibilities.

In a more speculative vein, one might even suggest that the fundamental difference between these models and the values they embody is a difference in the conception of the social order presupposed in these two models. One might suggest that what lies behind the commercial contract model is a picture of a social order in which individuals are free to pursue their own autonomously-chosen goals along the paths they see most plausible so long as they do not infringe upon the rights of others. One might suggest that what lies behind the status model is a conception of the social order in which individuals are obliged to accept responsibility for the well-being of themselves and others in accordance with some objective conception of what constitutes individual well-being. But this is, of course, merely a speculation.

What we have said so far makes it relatively easy to see both the strengths and weaknesses of the commercial contract model. Most of us share with this model a commitment to the ideals of freedom and autonomy. Because we do, we find certain features of this model extremely attractive. The theme that a patient should be able to choose whether to be treated or not and the theme that a physician should be able to structure the medical treatment he provides in accordance with his vision of proper medicine providing that the patient consents to that vision must surely seem attractive to all those who understand and appreciate the ideal that rational adult human beings should be free to structure their lives in accordance with their goals, perceptions, and plans. At the same time, most of us share the ideal that human beings bear some responsibility for the well-being of their fellow human beings. Because we do, we may be troubled by other themes of this model including the theme that physicians can refuse to treat prospective patients in emergency situations and the theme that society is not responsible to pay for medical care for the indigent.

What we have said so far also makes it easy to understand both the strengths and weaknesses of the status model. Indeed, the strengths of the status model (e.g., its insistence upon social responsibility for bearing the cost of medical care and its insistence upon the physician's responsibility for treating in emergencies) are precisely in those areas which were the weaknesses of the commercial contract model, and the weaknesses of the status model (e.g., its requiring the patient to seek treatment even when the patient prefers not to and its requiring physicians to treat) are precisely in those areas which are the strengths of the commercial contract model.

Let me again suggest, as a speculative suggestion, that much of the current uneasiness in the medicine/law relation is precisely due to the tensions caused by our adherence to various aspects of each of these models. It is not easy to simply pick bits and pieces from each model and put them together without feeling a sense of tension.

What shall we do as we face this impasse? There are two options that seem open: (1) we may view this conflict as part of the broader problem of freedom vs. paternalism. We may say that what we have here is one more case in which we have to balance the claims of freedom and autonomy against the claims of individual and social paternalistic responsibility for the well-being of others. If we say this, then we are likely to conclude that no resolution of this conflict is possible until a viable resolution of this broader conflict has been reached. (2) There is, however, another way of looking at this conflict. It begins by noticing that there is one feature which both models share in common. Both models try to treat the patient-physician relation by analogy to other relations outside the sphere of medicine. The commercial contract model tries to pattern the patient-physician relation on the relation that holds in commerce between the provider of a service and his customers. The status model tries to picture the patient-physician relation on the model of a stranger coming to the help of another stranger in an emergency. Both models, it might be suggested, fail because of this common feature, and a proper resolution of this conflict requires taking its point of departure from the *sui generis* nature of the patient-physician relation.

Although I cannot argue the case here, I would at least like to suggest that this second approach holds out the most promise for the development of an adequate legal model of the relation between patient and physician and between both of them and the society in which they live. The physician is not a stranger who occasionally comes to the aid of others; he is a professional who is devoting his life to the helping of others, but who has his own goals and plans which deserve respect. The patient is not just the purchaser of service in a market. All too often, he is a frightened, insecure, and pained person who needs friendship and responsible caring and not just commerce. Society is not just a passive umpire; it is also not an all-embracing dictator. Only when we see the patient-physician relation and its social context in its *sui generis* nature will we begin to develop an alternative and preferable legal model for the patient-physician relation.

Center for Ethics, Medicine, and Public Issues,
Baylor College of Medicine, Houston, Texas

NOTES

[1] The use of this name may (and is designed to) remind the reader of the relevant distinction drawn by Sir Henry Maine in his classic work [7], although we do not accept his view that progressive societies involve a move from status to contract.

[2] The first of these statutes was, apparently, the 1959 Amendment to Section 2144 of the California Business and Professions Code which read: "No person licensed under this chapter, who in good faith renders emergency care at the scene of the emergency, shall be liable for any civil action for damages as a result of any acts or omissions by such person in rendering the emergency care."

[3] The other major source of such cases involves patients' refusing artificial life-prolonging measures. One important case is [3], which certainly casts doubt upon New Jersey's continued adherence to the opinion of Chief Justice Weintraub quoted below. In other states, statutes (e.g., California Health and Safety Code, Section 7185, Idaho Code Section 39–4501, New Mexico Stats Ann. Section 24–7–1) have allowed such patients to refuse further treatment.

[4] Among such decisions are [2] and [4].

[5] Further evidence for the continued operation of this model is to be found in the material collected in [9].

[6] Although, to be sure, these requirements and acts go beyond just protecting against fraud. On this point, see ([1], Chapters 12–14).

[7] See the commentaries on *Yoreh Deah*, Chapter 336.

[8] A good place to look at such a discussion is in R. Moshe Feinstein's *Iggerot Moshe, Yoreh Deah* 58.

BIBLIOGRAPHY

1. Brody, B.: 1982, *Ethics and Its Applications*, Harcourt, Brace, Jovanovich, New York.
2. *In Re Brooks' Estate*, 1965, 205 N.E. 2nd 435.
3. *In the Matter of Karen Quinlan*, 1976, 355 A. 2nd 647.
4. *In Re Osborne*, 1972, 294 A. 2nd 372.

5. *J. F. K. Memorial Hosptial v. Heston*, 1971, 279 A. 2nd 670.
6. *Kennedy v. Parrot*, 1956, 90 S.E. 2nd 754.
7. Maine, Sir Henry: 1861 (1954), *Ancient Law*, Oxford University Press, New York.
8. *Mohr v. Williams*, 1905, 104, N.W. 12.
9. Rosoff, A. J.: 1981, *Informed Consent*, Aspen Systems Corporation, Germantown, Maryland.

PATRICIA D. WHITE

THE COMMON LAW AS A MODEL OF THE PATIENT-PHYSICIAN RELATIONSHIP: A RESPONSE TO PROFESSOR BRODY

If as he suggests at the outset, and as I suspect, the point of Professor Brody's essay is to make the claim that neither the common law nor Judaic law provides us with a model which fully and adequately captures the morality of the patient-physician relationship, then, although he importantly misdescribes the common law model, he is doubtless right. But if, as his conclusion suggests, his point is the further one that the law ought somehow to be reformed so as to reflect accurately the morality of the patient-physician relationship, then his claim is both controversial and unestablished. Since I doubt that Brody really intends to put forth the latter view either as stated or in a form generalized to include other sorts of social relationship, I propose to concentrate my remarks on the set of issues which are entailed by the first claim and largely to ignore the important jurisprudential questions raised by the second.

Brody sets forth two very different models of the patient-physician relationship – models which he says underlie its treatment in the common law and in Judaic law respectively. Testing these models by comparing them to his own perceptions of the morality of the relationship, Brody concludes that neither model gives a sufficient picture of its normative aspects. Although Brody does not do much to spell out the moral constraints which he thinks ought to characterize the patient-physician relationship, his method of proceeding suggests that it might be instructive to look briefly at the relationship itself to see what sorts of constraints might be appropriate.

If we think of the patient-physician relationship as one in which each party has obligations *to* the other we can begin to see some useful distinctions. Thus, to say that someone is obligated to become a patient does not entail that he is obligated to any particular physician to become his patient.[1] Even if a person is obligated as Brody says we are under Judaic law, always to seek medical attention when he is ill, it does not follow that before he becomes a patient he is thereby obligated in any way to the physician who treats him.[2] By contrast, however, if we say that a physician is obligated by virtue of his training and knowledge, whenever he is able, to help people who need medical attention, it often would follow that the physician would thereby be obligated to the person who subsequently became his patient. The

Earl E. Shelp (ed.), The Clinical Encounter, 133–139.

first of these two observations suggests that a patient's obligations *qua* patient to a physician do not arise until a patient-physician relationship has been established – until the person has in fact become a patient of a particular physician. The second indicates that there is a lack of symmetry to the relationship – physicians may be obligated to a specific potential patient even if the potential patient is not obligated to the physician.

Even if the obligations of a patient *qua* patient to his physician do not arise until their patient-physician relationship has been established, they would at least include any legally permissible contractual obligations that the patient undertook at the time he entered into the relationship (normally these would relate to the terms of payment). They might, of course, also include other obligations – perhaps, for example, an obligation to disclose and describe his symptoms to the physician as fully and accurately as possible.[3]

The obligations of a physician *qua* physician to his patients and potential patients are more numerous and difficult to characterize. They have their genesis in several sources. Once a patient-physician relationship has been established, the physician has at least a *prima facie* obligation to the patient to satisfy any specific representations that he has made concerning the treatment (its content, its effect, its cost, etc.) that he has undertaken to give. However, tempering these representations as well as the entire scope of the relationship are the special obligations imposed by law, by the canons of the medical profession, and perhaps by morality specifically on the behavior of physicians toward their patients in addition to all of the obligations, from whatever source, that the physician has, independently of his status as a physician, to the patient. Many, if not most, of these obligations cannot be waived by the patient and cannot be contracted away by the physician.

Such obligations as a physician *qua* physician may have to a potential patient cannot, of course, have arisen as the result of any representation made by the physician to the potential patient himself,[4] but they are nonetheless tempered by other obligations – some peculiar to physicians and some which apply to people generally. And although it is surely true that most people in such physically dire straits as we can imagine most potential patients to be *would* consent to medical treatment if they were in a position to, it is equally true that some might not and that neither the obligation of the physician to treat such a person nor the justification of his having done so can be grounded in any notion of the potential patient's tacit consent.

Sketchy though it is, it should be plain that this description of the patient-physician relationship is inconsistent with the models of that relationship which Brody ascribes to the common law and Judaic law respectively. Since I have no reason to believe that Brody would find my description unacceptable as a rough beginning, it is easy to see why he concludes that neither legal model satisfactorily captures the morality of the patient-physician relationship.

In fact, however, the common law comes far closer to mirroring my description than Brody gives it credit for. Brody pictures a common law which views the patient-physician relation as essentially a contractual one and, accordingly, he constructs a model in which the patient and the physician have freely entered into their relationship on mutually agreed upon terms. His model accounts for the obvious instances where the patient has been given emergency treatment, and thus had no chance to agree to any terms whatever, by allowing that sometimes the patient's consent is hypothetical. Presumably, on this model, each party has recourse against the other when the other fails in some way to live up to the agreed upon terms.[5] A typical action for medical malpractice where a patient sues his physician for negligently causing him some injury, could then be explained by Brody as representing the patient's claim that the physician failed to keep his (usually tacit) representation that he would perform his job competently. But this explanation, and indeed the whole contractual model, could not easily account for a successful negligence suit against a physician who had expressly contracted not to be liable for any damages that resulted from his negligence. What is missing is any acknowledgement of the standards that the common law recognizes as governing the patient-physician relationship and its formation quite independently of its contractual basis. These standards are to some fair degree established and enforced through the mechanism of tort law.

In the usual suit for negligently caused harm, the plaintiff must show, among other things, that the defendant acted less carefully than a reasonable person in the defendant's position should have been expected to act under the circumstances. Generally, the standard of care which is held to be appropriate is determined on a case by case basis by the trier of fact – either the judge or a jury.

A medical malpractice suit is frequently brought as a species of a suit for negligently caused harm. The standard of care to his patient to which a physician is held is of course affected by the fact that he is a physician, but it is also affected by local practices and reasonable expectations. Thus in treating a spinal fracture, an orthopedic surgeon at a sophisticated medical school is likely to be held to a higher standard than is a general practitioner who is the only doctor in a small town. This likelihood remains even if the patient was himself not sophisticated enough to realize that the standard of care one can reasonably expect will vary from place to place. The determinations of what the appropriate standard is and whether, in his treatment of a patient, the physician failed to live up to it are, in a malpractice suit, made by the trier of fact. Are these determinations, as they would be under Brody's model of the common law, normally made on the basis of the agreed upon (even if only tacitly) expectations of the parties, or are they often made independently of any such agreement?

It is clear that most doctor-patient relationships are not initiated by the signing of a contract. Nor are they usually marked by the explicit discussion by the parties of their respective responsibilities. Any contractual or quasi-contractual relationship between the two will often be largely tacitly established. Even if it is based on a tacit agreement, the obligation of a patient to pay a physician who has treated him is nonetheless conceived by the common law as being contractual in nature and a physician's suit for fees due him would likely be brought as an action for breach of contract. This is so even if the parties had rather different ideas of what the amount of the fee was to be. In the absence of any evidence of an explicit agreement between them, a court might well base its determination of the amount due the physician on *its* sense of the usual and customary amount paid for comparable services in that location. It would seem in such circumstances that the patient is being said to have contracted tacitly to pay the relevant standard fee.

The physician's obligations to the patient are harder to account for as a matter of tacit agreement. In most cases, the physician may well be said to have undertaken to conform to appropriate professional standards in his treatment of the patient. The patient may or may not have any understanding of what those standards entail, but he is likely to understand,[6] roughly at least, that the physician is undertaking to act professionally. However, although in his treatment of a patient the physician may expressly agree to do more than the relevant professional standards would require of him, he cannot, even with the explicit agreement of the patient, protect himself from liability in a malpractice suit by electing not to have his performance judged by the law by those standards. To so contract would itself be in violation of the minimum standards to which he must adhere and the agreement would have no legal effect. In an action for malpractice a court must apply whatever it determines to be the relevant standard of due care without regard to any contractual attempt by the parties to lessen the physician's responsibility.[7] This fact should at least cast doubt on the adequacy of Brody's "commercial contract" model as a description of the common law.

Courts are seldom as self-conscious about the conceptual roots of the law and of their decisions as philosophers might wish them to be, and so we cannot simply look to the recorded cases to see if judges conceive of the patient-physician relationship under the common law as contractual in Brody's sense. But the issue of whether the relationship does have a contractual basis in the standard legal sense does sometimes explicitly arise in the context of deciding whether the statute of limitations established by a state for suit brought for breach of contract or the frequently shorter one established for suits brought in tort applies. A particularly thoughtful analysis of this issue was offered by Judge Minor Wisdom in *Kozan v. Comstock* [1].

The serious question before us is determining which prescriptive period is applicable. If the action sounds in contract the prescriptive period is ten years [in Louisiana]. If it is a tort action then the suit is barred after one year. . . . There is a substantial difference between a physician suing an employer of persons whom the physician treated for his fee and a patient suing a physician for injuries suffered. In the first situation the establishment of a contractual relationship is essential, for otherwise there is no basis upon which a third party can be sued. In a suit by a patient against a physician for injuries suffered it is not essential that a contractual relationship exist. The duty owed by the physician to the patient arises as a matter of law and the physician is liable for a breach of this duty. . . . Decisions in other states may be divided into three categories. Some states hold that the action sounds in tort and is subject to the shorter statute of limitations applicable to tort actions. Other states hold that it may be brought in tort on implied contract. A third approach is represented by those states which have a specific statute of limitations that applies to malpractice suits against physicians and surgeons. . . . The causes of action in tort and in breach of contract for malpractice are dissimilar as to theory, proof, and recoverable damages. . . . It is the nature of the duty breached that should determine whether the action is in tort or in contract. To determine the duty one must examine the patient-physician relationship. It is true that usually a consensual relationship exists and the physician agrees impliedly to treat the patient in a proper manner. Thus, a malpractice suit is inextricably bound up with the idea of breach of implied contract. However, the patient-physician relationship, and the corresponding duty that is owed, is not one that is completely dependent upon a contract theory. . . . Even in . . . instances in which no contract is present the physician still owes a duty to the patient. The duty of care is imposed by law and is sometimes over and above any contractual duty. Certainly, a physician could not avoid liability for negligent conduct by having contracted not to be liable for negligence. The duty is owed in all cases, and a breach of this duty constitutes a tort. . . . This view that malpractice suits are tortious in nature probably represents the majority view. . . . We do not mean to say that there can never be a contractual action against a physician. . . . However, in the absence of a special warranty or contract, a malpractice suit against a physician is an action in tort and is subject to the limitation period for tort actions.

Although Judge Wisdom's analysis in *Kozan* is necessarily premised on the view that there is a fundamental distinction to be drawn between a suit in contract and a suit in tort and, as a matter of conceptual analysis, we need not be bound by that view, it seems nonetheless to point to the conclusion that the common law does conceive of and enforce standards of physician behavior independent of even the tacit agreement of the physician or the hypothetical agreement of his patient.

It would be misguided to conclude, however, that because the common law does impose independent standards on the behavior of physicians, it therefore adequately captures the morality of the patient-physician relationship. The standards of appropriate professional behavior which are enforced by tort law doubtless reflect to a large degree people's views of the moral standards which ought to govern a physician's actions. But there is little reason to imagine that they are identical and even less to suspect that they are the same as the moral standards themselves. Nor is there any reason to expect that the standards of the common law would be particularly instructive in an effort to construct a fully adequate moral model of the patient-physician relationship. On this point Brody is surely right.

Georgetown University Law Center,
Washington, D. C.

NOTES

[1] Nor would it normally even entail that he is obligated to become the patient of any particular physician, although it might if there were only one appropriate doctor available to him.

[2] The sort of general obligation to seek medical help when it is needed that Brody finds in Judaic law might, under some circumstances be construed as an obligation to do as the physician advises, but it seems unlikely ever to be an obligation owed *to* the physician (except in so far as it is an obligation owed to society and to its members and the physician is a member of society).

[3] Throughout this discussion I speak of obligations which are generated by the patient-physician relationship. Many, if not all, of these obligations could doubtless be subsumed under more general obligations (here, for example, the obligations to tell the truth and not to mislead), and thus, strictly speaking, are not triggered by the patient-physician relationship even if their character is determined by it. A complete analysis of the morality of the relationship would have to deal with this problem.

[4] Once a physician has made representations to someone which obligate him to act toward that person in the capacity of a physician, a patient-physician relationship would seem to have been established. It is not always obvious exactly when a potential patient becomes a patient.

[5] Because they commonly deal with legal obligation, lawyers tend to think of obligations as giving rise to causes of action if they are not kept. That habit of mind is actually useful in thinking about what the legal model of the patient-physician relationship might be.

[6] I am assuming here that the patient is a mentally competent adult.

[7] In unusual circumstances it is possible that the parties might in fact succeed in lowering the standards to which the physician must adhere. If the patient were fully informed and chose freely to agree that the physician not be liable if he acted negligently, a court might preclude him from later recovering in a malpractice suit if it found that his agreement constituted contributory negligence. This would not, however, be the same thing as its upholding the contract.

BIBLIOGRAPHY

1. *Kozan v. Comstock*, 270 F.2d 839, 842–845 (1959).

ISAAC FRANCK

JEWISH RELIGIOUS LAW AS A MODEL OF THE PATIENT-PHYSICIAN RELATIONSHIP: A COMMENT ON PROFESSOR BRODY'S ESSAY

My comments on Professor Brody's stimulating and provocative paper will not deal either with his sketch of the 'commercial contract' model of the patient-physician relation, or with the common-law treatment of this relation, a treatment that, according to Brody, "is captured by" the commercial contract model. This is not my assignment. Neither will I be concerned with any systematic, critical assessment of what Professor Brody calls the 'status model' of physician-patient relations. What I wish to examine is whether and to what extent the 'status model' in fact "captures the treatment of the relation in Judaic Law," as Brody claims, or fails to do so. The other principal objective of this commentary will be to amplify Brody's penetrating account of the ethics of physician-patient realtions in Jewish law, where such amplification seems needed, and also to comment briefly on the societal values and principles that are both assumed and implied by a full understanding of these relations.

Professor Brody renders a useful service in outlining five ways in which the status model captures the 'Judaic Law treatment' of the physician-patient relation, namely, (a) that both parties to the patient-physician relation are under obligation to enter the relation; (b) that both patient and physician enter the relation under terms set in part independently of their wishes, i.e., *the physician* – when and how he can treat, and how much he can charge, and *the patient* – what kind of treatment (risky or experimental) he can demand, accept, or refuse; (c) that the physician-patient relation is governed by these obligations and terms, and not based on mutual consent, e.g., no provision of questionable or dangerous treatments for higher fees, even if both patient and physician agree; (d) that society's main role is to insure that each party enters the relation when they are obliged to do so, and to insure the appropriate governing of the relationship; and (e) that society's role may entail its paying for the medical treatment, though the question remains whether the physician can demand that the community pay him for treating indigent patients.

Brody is right in pointing to *responsibility* as the fundamental value that lies behind the status model of the physician-patient relationship, as reflected in Jewish law, emphasizing as it does the threefold responsibility, namely, the

Earl E. Shelp (ed.), The Clinical Encounter, 141–148.

physician's to treat patients, the patient's to seek the best medical treatment, and society's to fund the medical treatment required. He is also right in his hesitant delineation of the conception of a social order that lies behind the status model of the patient-physician relation (as captured in Jewish law). In this social order persons are expected to accept responsibility for the well-being of themselves and others pursuant to "some objective conception of . . . well-being." But it is in Brody's reticence here, which leads him to suggest that the above "is . . . merely a speculation," that one discovers the need for substantial amplification of his interpretation of the patient-physician relation in Jewish law. Brody may well be justified in labelling as "merely a speculation" the conception of the kind of social order he suggests might lie behind the *status model* (and it is not my intention to discuss that here), but the attribution of the idea of this kind of social order can hardly be called speculative when what is under discussion is the authentic patient-physician model as conceived in *Judaic law*. In this respect, i.e., in the unequivocal, non-speculative character of its relation to this conception of a social order, and in at least two other respects that will be explicated below, lie some of the divergences of the patient-physician model in Jewish law from the status model as developed by Brody.

Having claimed categorically that the status model of the patient-physician relation "captures the treatment of the relation in Judaic law," and having expounded, with admirable learning, five ways in which the status model does in fact capture essential aspects of the treatment of the patient-physician relation in Jewish law, Brody proceeds to suggest the "strengths and weaknesses" of the status model (as he had also of the commercial contract model), differentiating them sharply from the strengths and weaknesses of the commercial contract model. However, there is one feature Brody discovers which both the commercial contract and the status patient-physician model share in common. Both these models, says Brody, "try to treat the patient-physician relation by analogy to other relations outside the sphere of medicine." So far as the status model is concerned, according to Brody, it "tries to picture the patient-physician relation on the model of a stranger coming to the help of another stranger in an emergency." Also, Brody claims this model fails (as does the commercial contract one) because it does not take "its point of departure from the *sui generis* nature of the patient-physician relation."

At the point of the above two characterizations, the suspicion begins to arise that while Brody's exposition of the ways in which the status model captures the treatment of the patient-physician relation in Judaic law is an interesting and instructive exercise, it could not have been intended by him to imply an identity or complete congruity between the two systems. For if indeed the *status model* tries to picture the patient-physician relation as that of a "stranger coming to the help of another stranger in an emergency," surely there is little resemblance here to the relation in *Jewish law* when it is read in its full authenticity.

Neither does the treatment of perception of the patient-physician relation in Jewish law ever follow "by analogy . . . other relations outside the sphere of medicine," thus being guilty of departing from the *sui generis* nature of the patient-physician relation. That this may erroneously appear to be the case is a result of the fact that Jewish law on these matters today is not perceived in its full authenticity. Our perception of Jewish law on the morality of the patient-physician relation is refracted through the prism of Western common law, grounded as it is in its social philosophies of absolute autonomy and freedom for interacting individuals. In a society and system so conceived, the philosophy of medicine that is unavoidably envisioned seems to entail perceiving the patient-physician relation as a *dyadic* relation, and transactional in character.

Carefully examined, and seen in its full authenticity, Jewish law treats the patient-physician relation as unique, as *sui generis*, not modelled on any "other relations outside the sphere of medicine." Moreover, as conceived within Jewish law, there always is a third party in the medical encounter, and thus the patient-physician relation is, paradoxical though it may sound, a *triadic* relation. In the medical encounter, the ever present third party is God, who is the *rophe holim* [Daily Prayer Book], the healer of the sick, who makes it incumbent (as expounded by Brody) upon the physician to heal and upon the patient to seek treatment, making both patient and physician God's surrogates, though in different ways, so that the physician cannot not treat and the patient cannot not seek treatment. What we have here is a *covenantal*, triadic relation between patient(s), physician(s), and God (or God's *Torah*), to collaborate with God in the process of continuing creation to mend the imperfections in the world and in men. This is more than a metaphorical notion. The point is a theological one and a metaphysical one, with clinical, legal, and social implications. It is charmingly conveyed in a well-known second century *Midrash*:

It is told of Rabbi Ishmael and Rabbi Akiba that they were walking in the streets of Jerusalem accompanied by another man. They were met by a sick man who said to them: "Masters, how can I be healed?" They said to him: "Take such and such [a natural medication] until you recover." Then the man who accompanied them said to them: "Who afflicted him with his ailment?" They said: "The Holy One, blessed be He." To which he replied: "You have gotten yourselves involved in a matter which is of no concern to you! He [God] afflicted him, and you seek to cure him?!" Whereupon Rabbi Ishmael and Rabbi Akiba asked him: "What is your occupation?" The man replied: "I am a farmer, and here is the sickle in my hand." Then they asked him: "And who created the vineyard?" He replied: "The Holy One, blessed be He!" Then the sages said: "And haven't you gotten yourself involved in a matter that is not your concern? God created the vineyard and you eat its fruit!" Then he replied: "Do you not see the sickle in my hand? If I did not go out and plow, and cut down the brambles, and weed, and fertilize, the vineyard would produce nothing." They then said to him: "Foolish man that you are! Have you not heard what is written [in the Psalms]: 'Man's days are as grass. . .?' Even as the tree does not produce fruit if it is not fertilized, weeded, and plowed, and even as the tree that produces fruit cannot live but dies if it is not given water, so is it in the case of the human body. The body is the tree, the fertilizer is the medication, and the farmer is the physician" [*Midrash Shmuel*, Chapter 4].[1]

God did not create a perfect world, and the work of Divine creation still continues. Divine creation is an ongoing, continuing process, in which man collaborates as *shoutafo shel hakadosh barukh hu*, God's partner, to assist God in trying to correct defects and mend imperfections in this imperfect world. The Daily Prayer Book exalts God "Who in His goodness every day continually renews the work of creation." A person's illness is not a *private* matter. It is an imbalance, an imperfection in the community, a lack, a privative condition that needs mending. The relationship between the physician and the ill person in this system is *sui generis*. Unlike any other profession, the practitioner of medicine is *commanded* to practice his craft, as a *moral imperative*. It is an imperative not only (as pointed out by Brody) in order to avoid violating the prohibition against "standing by the blood of your neighbor." It is also an imperative because, even as a lost object must be returned to its owner [Deuteronomy, XXII:2], so the ill person must have his healthy body, or his health, which he has lost, returned to him [*Sanhedrin*, 73a].[2] In the rectification of this privative state, achievable through the triadic collaboration between the three parties, the patient, the physician, and God, the Supreme Healer, the special role of the physician is that of God's emissary (*sh'luha d'rahamana*) ([2], p. 19), who cannot evade that role:

The Torah gave permission to the physician to heal, and it (i.e., his doing healing) is a precept (*mitzvah*) . . . and if he refrains (from doing healing) he is a shedder of blood ([*Shulhan Arukh, Yoreh Deah*, 336:1]).

Moreover, in this triadic convenantal relation, the physician is not free to discontinue treatment and abandon his patient, even if he believes that the case is a hopeless one. This is illustrated in a didactic tale out of nineteenth century eastern Europe, where Western influences had already made inroads:

A story is told of the 19th century Polish scholar, popularly known as Reb Eisel Charif. The venerable Rabbi was afflicted with a severe illness and was attended by an eminent specialist. As the disease progressed beyond hope of cure, the physician informed the Rabbi's family of the gravity of the situation. He also informed them that he therefore felt justified in withdrawing from the case. The doctor's grave prognosis notwithstanding, Reb Eisel Charif recovered completely. Some time later, the physician chanced to come upon the Rabbi in the street. The Doctor stopped in his tracks in astonishment and exclaimed, "Rabbi, have you come back from the other world?" The Rabbi responded, "You are indeed correct, I *have* returned from the other world. Moreover, I did you a great favor while I was there. An angel ushered me in to a large chamber. At the far end of the room was a door and lined up in front of the door were a large number of well-dressed, dignified and intelligent-looking men. These men were proceeding through the doorway in a single file. I asked the angel who these men were and where the door led. He informed me that the door was the entrance to the netherworld and that the men passing through those portals were those of whom the *Mishnah* says, 'The best of physicians merits *Gehinnom*.' Much to my surprise, I noticed that you too were standing in the line about to proceed through the door. I immediately approached the angel and told him: 'Remove that man immediately! He is no doctor. He does not treat patients; he abandons them!' " [1].[3]

The unique, *sui generis* character of the patient-physician relation in the medical encounter within the framework of Jewish law is made additionally evident when we amplify the legal and moral doctrines, and also scrutinize the practices, in the matter of payment to the physician for his services. On the one hand, as Brody points out, ordinarily it is permissible for the physician to charge the patients he treats, and the patient is expected to pay; and there are extensive Talmudic and Rabbinic discussions and legal decisions on medical expenses, doctors' fees, and the payment of doctors' bills. On the other hand, the fact is that there was always an expectation that the physician will not charge the poor for medical treatment. The puzzling Talmudic saying, "The best of physicians is destined for hell"[4] is explained by at least two standard, classical commentators as no doubt referring to the physician "who is in position to treat the poor but does not do it" [*Tosfot Yom Tov,* on *Mishnah Kiddushin* IV:14].[5] In his extensive ethical will to his son, Rabbi Yehudah Ibn Tibbon (ca. 1120–1190), himself a physician, admonished his son, Rabbi Samuel Ibn Tibbon (1150–1230), also a physician, thus: "My son, turn a welcoming face to all human beings and visit their sick, and let your tongue be a source of healing for them, and though you accept payment from the rich, *heal the poor gratuitously*" ([3], p. 55). However, as Jewish communities grew in size during later medieval and early modern centuries, physicians did not always remember the triadic convenantal relationship within which they were engaged by virtue of the physician's medical-ethical identity, and consequently Jewish communities and their religious courts sometimes had to compel the doctor to give his services to the poor without charge. The idea that this is part of the *sui generis* patient-physician relation, not modelled by analogy on other relations outside of the medical encounter, is suggested by the fact that one can hardly think of any other practitioner-client relation in which the morality of the relation requires that the poor be ministered to gratuitously.

Another aspect of the payment of doctors' fees provides an even more striking illustration – our final illustration – of the insight into the *sui generis* patient-physician relation in Jewish law, and thus of the deviation of Jewish law from what Brody calls the status model. Professor Brody quotes from Epstein's *Arukh Hashulhan*, that so far as physicians' fees are concerned,

> the *Tur* (13–14th century Code by Jacob ben Asher) writes in the name of Nachmanides (1193–1270) that one can be compensated for the time and effort expended but not for the study one has first undergone . . . as God says about doing good deeds, much as I act without a fee, so should you act without a fee [*Yoreh Deah* 336:2–3].

This is a remarkable legal-ethical doctrine. It is reaffirmed categorically in the *Shulhan Arukh*, the authoritative and standard Code compiled by Joseph Karo (1488–1575), and is underscored in numerous subsequent *Shulhan Arukh* commentaries, including *Turei Zahav, Ba'er Hetev*, Rabbi Elijah (Gaon) of Vilna, and others. The physician may accept payment for his *effort* and *time*, but he is forbidden to accept remuneration for his *studies*, his *wisdom*, his *knowledge*, or for his *instruction* to the patient. In the first place, his giving these to the patient and thus helping to heal him, is tantamount to restoring to the patient his health or his body *which he had lost* [Deuteronomy, XXII:2]. This is the fulfillment of a Biblical commandment or precept, and just as God acts gratuitously, so should the physician act gratuitously when he fulfills a precept. However, there is implicit here an even more significant *imitatio dei*, profoundly imbedded in the nature of the medical encounter, and in the *sui generis* conception of the triadic patient-physician relation within Jewish law, as expounded above. *God* is the ultimate healer, and the physician is God's emissary, God's surrogate, indeed, God's collaborator. The uniquely *sui generis* element in the patient-physician medical encounter is not the physical effort, the travel, the time brought by the physician to the encounter, but rather the medical knowledge, wisdom, instruction with which he has been endowed by the Supreme Physician. In a significant sense this wisdom and knowledge do not belong to him. And just as the Supreme Physician brings *His healing* into the triadic medical encounter *without fee*, so must the human physician also bring his healing knowledge and wisdom and instruction into it without fee. The relationship here is, in the theory, *sui generis*, and no other relations outside the sphere of medicine are quite analogous to it.

It seems to me that these explications of some of the ways in which the model in Jewish law, as I understand it, of the ethics of the patient-physician relation diverges from the status model as explicated by Professor Brody, are in fact amplifications of Brody's account of this subject. On the other hand, I believe that the comments I have presented above about the ethics of Judaic law on these relations tend to give strong reinforcement to Brody's answer on what he calls "claim five," namely, that "society's role may entail its paying for medical treatment," and for the very reasons given by him. As to the question whether the physician who treats the poor can demand that the community bear the cost, there is of course a negative answer contained in my earlier comments, if what is meant is whether such a physician has a moral right to demand payment. But if, on the other hand, what is asked is whether within a framework of Jewish ethics and law a community should establish community-based and community-financed medical services that would provide adequate medical care for the poor, and employ physicians for that purpose, my answer would be that in an imperfect world, in which physicians do not and did not always live up to the high ethical expectations of their noble calling, such a communal health system would be both permissible and desirable. In fact, during the long centuries when the political conditions under which Jews were compelled to live necessitated autonomous Jewish communities, essentially self-governing in all but matters of political authority, Jewish community systems of health services existed in a number of cities and towns.

I now want to conclude with a brief comment on basic values and on concepts of social order. In the course of expounding some of my understanding of the interpretation of the patient-physician relation in Jewish law, I referred once or twice to the 'full authenticity' of Jewish law. I indicated or implied that it is from this 'full authenticity' of Jewish law that I draw my interpretation. In using that phrase I had in mind an amalgam of the mainstream of legal-ethical doctrine in the succession of Biblical, Talmudic, Rabbinic, and philosophical Jewish writing through the centuries. I had in mind as well the distillation of the collective ethico-legal experience of autonomous Jewish communities in history. In such a fully authentic system, undiluted by admixtures of the commercial-contract elements out of Western culture, it is of course clear, as Brody suggests, that the value of *responsibility* in the sense in which he develops the term, is a basic underlying value. Free medical treatment for the poor, and community funded medical services are readily available examples out of Jewish history, and were expressions of the essential understanding of the *sui generis* character of the medical encounter. This sense is still very much alive today in the free, open, and voluntary Jewish communities in democratic Western countries, with their intricate networks of social service institutions and health-serving institutions the most prominent among them.

Whether a society in which the equivalent of the 'full authenticity' of the Judaic legal-ethical system governing the morality of the medical encounter is a possibility within a free democratic state; and whether a voluntary Jewish community existing within a free democratic state can ever recapture the 'full authenticity' of the Judaic legal-ethical treatment of the *sui generis* medical encounter; or whether it could ever be recaptured in the State of Israel; or whether a theocracy would have to be the necessary condition for the realization of any or all of these possibilities, is an interesting subject for speculation.

Kennedy Institute of Ethics and School of Medicine,
Georgetown University, Washington, D.C.

NOTES

1. My translation.

2 See also, *Shulhan Arukh, Yoreh Deah*, on 336, Commentaries *Turei Zahav, Ba'er Hetev*, and *Hag'ro* (Rabbi Elijah of Vilna).

3 See the following Note (4) on "The best of physicians is destined for *Gehinnom*."

4 This Mishnaic saying, which occurs in *Mishnah Kiddushin* IV, 14, is generally viewed as something of a puzzle when it is seen in the light of the consistently favorable and even venerating attitude to physicians in Talmudic and Rabbinic literature, and commentators have felt prompted to explain it away. One of these commentators, *Tiferet Israel (Yavin)*, protests that this saying "is not intended as a disparagement to the physician, but rather praise for the skilled physician ... it refers only to one who *considers himself* the most skilled among physicians, because in this arrogance of his he relies on his own doubtful knowledge and does not consult his colleagues."

5 Also *Rashi*, on the same *Mishnah*, in Babylonian Talmud, *Kiddushin*, 82a.

BIBLIOGRAPHY

1. Bleich, J. D.: 1976, 'Karen Ann Quinlan: A Torah Perspective', *Jewish Life* (Winter), pp. 13–20.
2. Wishlitski, L. (ed.): 1976, *Madrikh R'fuee L'fee Hamassoret Ha'yehudit* (Medical Guide in Accordance with Jewish Tradition), Institute for Medicine and Judaism, Jerusalem.
3. Zahavi, J. T. (ed.): 1952, *Mivhar Hamahshavah V'hamussar Ba'yahadut: Me'rav Sa'adia Gaon ad Yameinu* (A Selection of Thought and Ethics in Judaism: From the Time of Saadia Gaon to Our Own Day), Abraham Tzioni, Publishers, Tel Aviv.

BARUCH A. BRODY

RESPONSE TO FRANCK AND WHITE

The insightful and helpful remarks of Professors Franck and White raise a great many questions, and I obviously cannot deal with all of them in this brief response. I shall focus on their claims that neither the common law nor Judaic law rules governing the patient-physician relation are captured by either the commercial contract or the status model.

Let me begin by agreeing with Professor Franck that in the Judaic conception of the patient-physician relation, God is certainly present as a third party who helps structure that relation. In truth, that point can also be made about any other legitimate relation, since Judaism sees God as present in and as helping to structure all such relations. So while I agree with his point, I do not see that it helps establish his broader point that Judaism treats that relation in a *sui generis* fashion.

That leads me to my second set of comments. I do not see that any of his other points establish that Judaism treats that relation in a *sui generis* fashion. Franck thinks that the treatment of fees for physicians establishes this point. But the authorities he cites develop their point by comparing the physician's fees to the fees charged by others who perform *mitzvot* (commandments) in a skilled fashion. Again, the very commandment of the physician to heal is the same commandment that applies to any stranger to come to the need of others whose life is threatened. So while Franck's points are correct, I do not see how they are supposed to establish the truth of his main contention, that the patient-physician relation is treated by Judaism in a *sui generis* fashion.

I turn then to Professor White's paper. I think that the early part of it is very helpful in at least raising some of the questions which need to be considered if we ever are to develop an adequate model for the patient-physician relation. I want to focus on the second part of her paper, where she argues that too much of the standards for treatment and care are set independently of any explicit contracting between the two parties for the contract model to be helpful. Against her observations, I would only make three points: (a) most contracts involve large numbers of terms set by common practices and understood standards; (b) while there are some terms (e.g., release of liability for negligence) which we will not allow physicians

Earl E. Shelp (ed.), The Clinical Encounter, 149–150.

and patients to contract for, this is also true in other areas where contracts are made (think of contracts against public policy, unconscionable contracts, etc.); (c) it is still within the power of both parties to structure by explicit contract many of the terms of the relation. I cannot see then that Professor White has shown that the common law treats the patient-physician relation all that differently from the way it treats other contractual relations.

One final point: that a relation is based upon contract only means that the duties it creates are created by contract. Negligence in the performance often can (and probably should) still be treated by separate tort rules, including statute of limitation rules. Breach of contract is one thing; misfeasance in performance is another.

Center for Ethics, Medicine, and Public Issues,
Baylor College of Medicine, Houston, Texas

SECTION III

CONCEPTUAL AND THEORETICAL ANALYSES

EDMUND D. PELLEGRINO

THE HEALING RELATIONSHIP: THE ARCHITECTONICS OF CLINICAL MEDICINE

> Thus the Art of Medicine rules and orders the Art of the chemist because health with which medicine is concerned is the end of all the medications prepared by the chemist. The arts that rule the others are called Architectonic, as being the ruling arts. (Thomas Aquinas, *Summa Contra Gentiles*, Bk. 1, Ch. 1.)

INTRODUCTION

In his devastating novel of the destructive force of human passion, the Spanish philosopher-novelist, Miguel de Unamuno y Jugo records this conversation between the painter, Abel Sanchez, and the doctor, Joaquin Monegro:

> – The habit of giving titles to paintings is peculiar to the literati; something like the doctor's habit of giving names to diseases they can't cure.
> – And whatever told you that the real purpose of medicine was to cure illnesses?
> – What is it, then?
> – Knowledge, a knowledge of disease. The end of all science is knowledge.
> – I had thought it was knowledge to cure. What use otherwise of having tasted the fruits of good and evil, if not to free ourselves of the evil?
> – And the end of art, what is it?
> – That is its own end; it contains its purpose. It is an object of beauty and that's enough ([6], p. 366).

In this brief exchange de Unamuno captures the central dilemma of modern medicine – What is it? What is it for? What knowledge does it need?

It is paradoxical, and frustrating, that after so many centuries we are still uncertain about what we mean by medicine, illness and health [14]. Yet, all our debates about medical practice, education, and ethics still must center on the unsettled question of the nature of medicine.

This is more than a trivial or contrived dilemma. What we think medicine is, determines what we think physicians ought to do, what they need to know, and what they should be taught. Each philosophy of medicine creates its own model of disease and health, of resource allocation and public policy.

Earl E. Shelp (ed.), The Clinical Encounter, 153–172.

Most critically it also shapes the relationships of physicians to patients and society. It makes a real difference therefore what philosophy we espouse. Unless we are clear about what medicine *is*, we risk deceiving ourselves, our patients and our society.

The resolution of the dilemma has never been more difficult or important. The capabilities of modern medicine, its frightening costs, its ethical challenges, and its influence on society and culture force us to grapple with the philosophical issues. That is why the philosophy of medicine has become an expanding and legitimate field of inquiry [21].[1] A wide range of competing philosophies is being advanced and must somehow be sorted out. At one extreme is Donald Seldin's restrictionist view that medicine is applied biology [28]; at the other, the global pretensions of the WHO definition [5]. In between, we find a spectrum of definitions – the bio-psychosocial model, the holistic concept, and theories grounded in medicine's social and cultural contexts or in some *a priori* definition of health.[2] Each depends upon some partial truth about modern medicine. Medicine has indeed advanced prodigiously because of its marriage with science; it has learned to benefit patients by a knowledge of psychobiology and psychosociology; it has improved human existence immeasurably by its preventive and public health contributions; it is learning how familial, social, and cultural contexts are implicated in the genesis of illness and its treatment.

The problem is not the denial of these obvious truths but rather, how much should be included under the rubric of medicine? How much is the physician's responsibility? How much belongs to other professions, to individuals, and society? How do we delimit what properly should be counted as medicine? We know we must practice some economy in our pretensions. Yet, we must also afford the patient what he or she needs, to be cured, healed, or helped.

What competing philosophies of medicine are seeking is an architectonics – some ordering or organizing principle whereby the boundaries of medicine can be determined. They seek it in a variety of places and ways. Some regard medicine primarily as a body of knowledge and confine it within the perimeters of a predetermined set of disciplines. Others define medicine in terms of its end or purpose or some predetermined concept of health for which medicine is the means. Still others would define medicine as the negotiated outcome of the physician-patient relationship.

My purpose, in this essay, is to sort out some of these theories, examine their logical and epistemological difficulties and place them in some orderly relationship with each other. I will also propose an organizing principle derived from an empirical and phenomenological examination of medicine as a human activity of a special and unique kind. My thesis is that it is in the nature of the healing relationship that we will find what is unique about clinical medicine, what is its immediate end and what will serve as its architectonic principle.

SOME THEORIES AND MODELS OF MEDICINE

Medicine as a Body of Knowledge

The commonest approach to a definition of medicine is to identify it with some body of knowledge useful in the treatment of illness. The dominant example of this genre is the so-called biomedical model, well articulated by Donald Seldin. Seldin defines medicine as biology applied to the cure and prevention of disease and the postponement of death. He specifically excludes social concerns, psychosocial elements and bioethics as beyond the boundaries of the physician's knowledge base, though he does not deny their importance in society [29]. Seldin summarizes his view this way:

> Medicine is a discipline which subserves a narrow but vital arena. It cannot bring happiness, prescribe the good life, or legislate morality. But it can bring to bear an increasingly powerful conceptual and technical framework for the mitigation of that type of human suffering rooted in biomedical derangements ([28], p. 40).

Seldin articulates clearly the dominant view of today's practitioners, teachers, and students. It has the virtue of limiting medicine to what it can do well, excluding what it does inadequately, and defining disease in terms of etiologies that are measurable and quantifiable. So attractive is the biomedical model that some investigations urge that it be the paradigm for psychopathology as well as somatic disease [15].

The limitations of this model are several. For one thing, it is flagrantly reductionistic, limiting medicine to biology, chemistry and physics. It suffers, therefore, from the logical and epistemological deficiencies of all reductionism [16]. It commits the error of circuitous reasoning – what we need for the proof is itself in need of proof by the settlement of the question we are asking. We cannot prove medicine is only biology unless we know first what medicine is – and that is the question.

The result is that the aims of medicine are determined by a preselected body of knowledge instead of medicine determining what knowledge it needs to achieve its ends. The error of reductionism is that is cannot logically exclude alternative reductions of medicine to applied sociology, psychology, or anthropology. Since these are admittedly elements of illness, why not assign them to the physicians and relegate the biological aspects to the biologists?

Even if we were to accept Seldin's reduction he does not tell us who is to provide the other kinds of knowledge needed to treat patients. How can those other dimensions be related to medicine? Who adjudicates the conflicts? Who decides what *should* be done for the patient? Since man is more than a biological being and his illness is more than biology, who manages the relationships between those different bodies of knowledge needed to heal? Seldin does not deny that human illness may rise in personal, familial, or social contexts or that they cannot be treated by biological means alone. What makes the patient 'whole' again – since this is what healing means?

The most serious empirical limitation of the biomedical model is its one-dimensionality, its abnegation of the complexity of the experience of illness and therefore of the complexity of healing those who are ill. François Jacob, the molecular biologist, shows how limited this one dimensionality can become:

> The one-dimensional sequence of bases in the genes determines in some way the production of the two dimensional cell layers that fold in a precise way to produce the three dimensional tissues and organs that give the organism its shape, its properties, and, as Seymour Benzer puts it, its four dimensional behavior ([13], p. 44).

Since disease and illness involve all four dimensions, it is hard to see how biology *qua* biology can be satisfactory as an explanatory principle to say nothing of a therapeutic one.

George Engel responds to the defects in the biomedical model by proposing one that includes biology as well as sociology and psychology and, he says, "will include the patient and his attributes as a person" ([9], p. 536). He (the doctor) must weigh the relative contributions of social and psychological as well as biological factors implicated in the patient's dysphoria and dysfunction as well as his decision to accept patienthood and with it the responsibility to cooperate in his own care. "It is the doctor's, not the patient's, responsibility to establish the nature of the problem and decide whether it is best handled in a medical framework." "Hence the physician's basic professional knowledge must span the social, psychological and biological for his decisions on behalf of the patient involve all three" ([8], p. 133).

Engel goes further and offers general systems theory as an explanatory model rather than biology. In this way, he proposes to resolve the controversies between reductionists and holists. Systems theory, he says rather expansively, should be the "... blueprint for research, a framework for teaching and a design for action in the real world of health care" ([8], p. 135). Having expanded the model of medicine he then confines it in a new mold – not of biology but of applied systems theory. The logical difficulties differ in degree but not in kind from Seldin's.

Engel's model has the virtue of including more of the realities of human illness, and vesting in the physician the task of fusing them in treating patients. It remedies some of the grosser deficiencies of the biomedical model but it still defines medicine in terms of a knowledge base, albeit a broader one. Indeed, so broad is this definition that it errs in the other direction, it leaves out very little. It would not take much analysis to stretch the biopsychosocial model to embrace almost as much as the much criticized WHO definition of health as complete physical, social, and emotional well-being.

Significantly, Engel does leave out one dimension – that is the spiritual, unless he includes it under the social and psychological. This is a dimension that the proponents of 'holistic' medicine feel is also a part of medicine and healing [1].[3] Since religion, or at least some stance with reference to the transcendental is part of the fabric of human existence, what justifies its exclusion?

Engel's theory along with Seldin's and the others that define medicine in terms of a knowledge base always ends up begging the question: What is medicine? Is it only the disconnected sum total of the kinds of knowledge it uses? If so, why is one combination of disciplines more pertinent than any other? The question can only be answered by recourse to something more fundamental, in the nature of medicine itself, something which shapes what knowledge is needed and for what purpose. Engel, for a brief moment, touches this question when he refers to what patients are seeking in their encounter with the physician [9]. He quickly leaves it to expatiate on the value of general systems theory as an organizing principle.

Systems theory may well provide a useful explanatory model for the interacting hierarchies of form and function that characterize human health and illness. But to make it, as Engel does, the 'blueprint', 'design', 'framework' of the whole of medicine is to trade on one kind of reductionism – the biological – for another. Granting the utility of such a step, it can become a shackle on medicine. The explanatory utility of a paradigm does not qualify it as a theory of medicine. It has always been perilous to transpose the method and language of one discipline beyond its operational boundaries. We need only mention the misleading and even deleterious effects of the appropriation of such concepts as Darwinian evolution, entropy, relativity, Freudian or Jungian metapsychology, or Wilsonian sociobiology by social, political, and adminstrative theorists to illustrate this point [24].

Neither Seldin's restrictionist nor Engel's expansionist theories are adequate as organizing principles. We are still left with the question: What is the nature of medicine, what knowledge does it need, what does it do?

Medicine Defined in Terms of Its End

An alternative approach to viewing medicine as a body of knowledge is to define it in terms of its end, its purpose, the terminus toward which medicine is directed as a human activity. The end then becomes the determining principle that defines what kind of knowledge medicine needs. Leon Kass is one of the few theorists of medicine who analyzes the nature of medicine this way – a genuinely philosophical way. He declares the healthy human being the only 'reasonable' goal of medicine. He rejects such goals as happiness, social adjustment, behavior modification or the simple prolongation of life. 'Health' is the end of medicine and not "... pleasure, happiness, civil peace and order, virtue, wisdom, and truth" ([14], p. 18).

Kass admits that he cannot define health precisely but he offers 'wholeness' and 'well-functioning' as approximations. Health is, he says, a "finite and natural norm," the "... well working of the organism as a whole, an activity of the human body in accordance with its specific excellence" ([14], p. 29). It is not only a somatic norm but a psychic one as well. Indeed, "our whole way of life" influences our health ([14], p. 31).

On Kass's view, attainment of the end of health is the doctor's only business. His role should be limited to the use of only the technology and knowledge that advance therapeutic purposes. It is the physician, not the consumer or the government that must define health care and the standards of practice. He questions expenditures for costly procedures like dialysis, and doubts the validity of such concepts as the right to health and the benefit of national health insurance. Health, while it is a good end in itself, is not the only ingredient of a good life. Those other ingredients are not the domain of medicine.

Having tried to contract the domain of medicine, Kass then opens it up again. He says health is everybody's business – not only the physician's. Medicine itself must become 'whole' again by attending to health maintenance as well as cure of illness. He urges more research in 'healthiness', disease prevention, and health maintenance and that these become the determinants of health policy.

Kass's analysis has the virtue of defining medicine philosophically in terms of the purpose for which it is ordained or acts. Physicians and patients do in fact seek health through the acts of medicine. That is why health has been recognized as the end of medicine since Hippocrates. The weakness of this theory lies in the indefiniteness of the definition of health. While Kass limits the physician to the attainment of health for his patient, his notion of health has such wavering boundaries that it could include even those things Kass specifically excludes. How, for example, can we place social adjustment or behavior modification outside medicine? Can emotional or psychological illness be ameliorated without some attention to social adjustment? Can prevention which requires a change in lifestyle be achieved without some modification of established behavior patterns? The doctor is enjoined to seek 'wholeness' and well function of psyche and soma. These ends require the same breadth of concern and knowledge that Engel includes in his theory.

Defining medicine by its end is more sound philosophically than defining it as a knowledge base. But unless the end itself can be delineated the boundaries balloon again when we try to realize that end in actual practice. Seldin's view at least has the virtue of specificity in its end – those illnesses susceptible to biological cure or amelioration. Furthermore, health is an abstract notion, an ideal towards which the doctor is supposed to work. As an ideal it is never fully attained. What is the doctor's function when health is not even relatively restorable, when the illness can only be ameliorated symptomatically, or contained? What is the end of medicine when even these limited goals are not attainable? Surely one of the 'ends' of medicine is to help the patient to cope with chronic illness or disability, or even to live the last days of life as comfortably and humanely as circumstances will allow.

Kass recognizes that responsibility for health is not solely medicine's. Everyone has some part in promoting his own health and society's. If this is so, we face the recurrent question: When is the pursuit of health the physician's duty and when not? Kass's theory never answers this question satisfactorily. It warns against using technology for 'non-medical ends' but this depends on defining 'medical' – something Kass does not do. As with Seldin and Engel, the central question remains.

Defining Medicine in the Patient-Physician Encounter

To obviate some of the difficulties of the knowledge and end-determined theories of medicine, one can approach the question more phenomenologically. Mark Siegler, for example, focuses his theory on the nature of the physician-patient relationship – on "... how clinical medicine works in the realities of daily practice" ([30], p. 631). He criticizes most theories because they are 'context-free' and deductive, to which he opposes his own concept which is 'context dependent' and inductive.

Siegler's major thesis is that "... a problem becomes one for clinical medicine only when the patient and doctor agree that it is one" ([30], p. 631). Siegler sees the definition of medicine as the result of a negotiated agreement between individual patients and physicians. He sees the negotiation in four stages: (1) The decision by the patient that he is ill; (2) the physician's application of the clinical method to answer the patient's question of whether he is sick or has a serious disease; (3) the doctor-patient 'accommodation' in which each decides whether or not to enter the relationship with the other; and (4) the doctor-patient 'relationship' in which an exchange of trust occurs and a stable and prolonged relationship is established ([30], pp. 631–641).

Except for the second stage, Siegler feels that theories of health and disease have relatively little influence on the definition of clinical medicine. Its nature is defined principally by the resolution of individual doctor-patient accommodations. In that accommodation Siegler clearly leaves the final determination to the physician. He says that "every patient presentation generates a claim to be heard by physicians" ([30], p. 643). But it is the physician who determines whether he will manage *this* problem in *this* patient.

Siegler's formulation has the signal advantage of seeking a definition of medicine in the phenomena of clinical medicine itself – indeed, in what is most characteristic of medicine, the encounter between physician and patient. His description of the four clinical 'moments' is sound so far as it goes. I too have used the term 'clinical moment' in defining what is unique about medicine – but in a rather different sense from Siegler's [17].

Unfortunately, Siegler does not really define clinical medicine. What he does is to describe the *process* whereby physician and patient agree to enter the clinical relationship, but not the essence of the relationship itself. The process of negotiation decides whether or not physician and patient will accept each other for what each claims to be – one claims to be sick, one claims to be able to heal. This would make medicine anything the physician and patient want it to be. Even if this were defensible in individual clinical encounters, it could not serve as a basis for public policy.

Siegler's four stages are propaedeutic to some end or purpose. It is what occurs after the decision is made that constitutes the clinical moment that sets medicine off from other activities. Doctor and patient may reject each other but presumably at some point each engages in some other therapeutic relationship – and what happens once the engagement occurs is what defines clinical medicine. The process of negotiation cannot exhaust what we mean by clinical medicine, hence it cannot *define* clinical medicine. This is not to deny the utility of Siegler's description of the preludes to clinical medicine but the prelude is not that which is essential to medicine as a human activity. Let us assume that the physician denies the patient's 'claim' to be ill. If the patient accepts this decision he has been relieved, reassured, and educated. He has been healed, cured, or helped. Something more than a mere negotiation has occurred. If the patient does not accept the denial of his claim to be ill, the reason he consulted the doctor remains unattended. He will seek another physician who does agree to go beyond the negotiation. In either case some end beyond negotiation of a claim about illness must be served.

The potential latitude of a definition by negotiation makes it difficult to translate into public policy, education, or ethics. The narrowest and the broadest interpretations would be equally tenable. On which of these bases do we establish the physician's ethical obligations, the education he should receive or the extent of what society should support as medical care?

Siegler's insistence on a definition based in the reality of clinical medicine itself is superior to the knowledge, or end-oriented definitions. It does not go far enough, however, into the central phenomenon of the physician-patient encounter. Since it leaves the ultimate definition to the physician who evaluates a patient's claim to be ill, the physician's philosophical conception of medicine is far more influential than Siegler allows.

Siegler shares with Seldin, Kass, and Engel the conviction that it is the physician's specific duty to define what is health, a view Whitbeck and others would contest. Whitbeck defines health as "... the psycho-physiological capacity to act or respond appropriately in a variety of situations" ([33], p. 611). She argues that medical expertise is not relevant to many health decisions and concludes that some of these decisions are best made by the person whose health is at stake. One might raise the same question about the definition of illness which is, after all, a subjective experience which the physician penetrates only partially and incompletely.

Fabrega takes the external definition even further, showing how social, personal, and cultural norms determine health, disease and illness [10]. Since medicine is part of the socio-cultural and ethnic fabric of societies it is related to the dominant characteristics of society. This view has merit for homogeneous cultures with univalent value systems. For multivalent systems like ours, the question remains: What is medicine? If it is defined externally, which of the many possible value systems should predominate?

Every society and era does indeed provide some system of succor for its members who, because of illness or trauma, cannot function in their accustomed social roles. The social and cultural definitions of health and illness may vary but the need for healing is a constant. It is in that constant fact that we should seek our organizing principle – the universal fact that humans become ill, and in that state seek and need help, healing and cure. However the cultural milieu may differ, this fact is common to all medical systems. Thus the immediate and practical end of medicine is an action taken in behalf of one who is ill, who has decided that his unease is significant enough to warrant a decision to seek medical help. Health is a more remote end. It is attainable only if the special human relationship of healing is a successful one. It is that relationship which provides the architectonics of medicine as medicine, and not health which may not even be attainable for many who seek help and healing.

THE HEALING RELATIONSHIP AS ARCHITECTONIC PRINCIPLE

No single theory can reconcile the conflicting definitions of medicine currently locked in debate. Granted their conceptual deficiencies, each derives from some truth about medical reality and thus has some heuristic value. But it is a long way from heuristics to a universal description of clinical realities. No single theory, including my own, can close this gap completely. I do believe, however, that the closer a theory comes to the phenomena of medicine as a relationship of persons the closer will it discern medicine's true nature.

I have offered elsewhere a model of medicine based in the clinical reality of the healing relationship [19]. I believe this is what distinguishes medicine from other human activities like baking bread, taking one's car to the mechanic, or buying clothes [27].[4] The healing relationship is a genuine architectonic principle for it shapes, directs, and bounds the physician's work, education, and ethics. In propounding this view, I can agree with selected aspects of the several theories I have discussed: with Seldin on the need for an economy of medical pretensions, with Engel on including the patient as well as the illness, with Siegler on the importance of clinical realities, with Kass on the need to define ends, and with Whitbeck on including the patient's assessment of what constitutes health. I am not proposing a 'please everybody' eclecticism; rather I acknowledge the validity of my colleagues' insights but differ substantially from most of them.

It may be objected that healing is not limited to medicine – and this is true. Psychologists, ministers, friends, and families can provide healing relationships. But they do so over a limited range of human need. Ordinarily the person who seeks extra medical healing does not consider himself sick – only troubled or anxious. Sickness implies embodiment, the distinctly human phenomenon of a conscious self in a lived body [21].[5] When a person experiences some disturbance in his accustomed state of balance between body, psyche and self he counts himself as sick. It is the fact of embodiment that creates the need for the physician. Only he can unravel the connections between the subjective experience of illness and its linkages to bodily function. Without denying the part others play, the physician comes closest to what healing means – to restore wholeness or, if this is not possible, to assist in striking some new balance between what the body imposes and the self aspires to.

What the sick person seeks is restoration to his or her definition of wholeness, that state in which Galen said we are unimpaired in doing the things we wish to do. This is as realistic a definition of health as we are likely ever to get. It depends as much on the patient's assessment as the physician's. If full restoration is not possible, then amelioration of suffering, adaptation, or coping with chronic or fatal illness become the ends of the healing relationship.

These are the ends specific to clinical medicine that distinguish it not only from other human activities, but also from other activities in which physicians may be engaged. Thus, medical science seeks the causes and pathogenetic mechanisms of disease and its end is knowledge. The end of preventive medicine seeks to promote the health of individuals and groups. Social medicine seeks the health of the entire population and its end is the public good. Physicians play important roles in all these activities but they can be accomplished without clinical skills. Only healing requires the physician as physician because its end cannot be achieved by others in society.

The end of clinical medicine, therefore, is a technically right and good healing action – one taken in the interest of a particular patient who is in need of, and seeks, help.[6] To help that person the physician must answer four questions – What is wrong ? What will it do to me? What can be done for me? What should be done for me? These are the clinical questions and they converge on the choice of an action. The moment of decision is the true clinical moment of truth, and in that moment what is most characteristic of medicine comes into existence.

Let us examine more closely what is involved in making a decision to act in a way that will be both technically right and good for *this* patient. This is far more than can be encompassed in the notion of medicine as applied biology or even as bio-psycho-sociology. Having a body of knowledge that may be pertinent to a clinical dilemma does not mean that the end of the healing relationship will be served. That knowledge must be directed to a specific practical end in a personal relationship in which one person is ill, vulnerable, and anxious and seeks the help of another who has knowledge that must be made to serve the interests of one who is ill.

Clinical medicine therefore only comes into existence at the moment of clinical truth, when a decision is taken and an action initiated to heal or help a particular patient. The true physician is the one who takes upon himself this responsibility, not as a negotiated task, but as an imperative built into the very nature of clinical medicine.

Medicine is therefore praxis and not theory. It contains its end within itself. Medical knowledge is incomplete as knowledge until it translates into action. It is also ethical knowledge because it is used for a good end. Although it uses science and art, medicine is closer in its moment of decision to the ancient definition of a virtue – an exercise in prudent action.

A technically right healing action must conform to several criteria. First, it must conform to the best available scientific information. Then it must draw on whatever sources are pertinent to the clinical state of this patient. This information must then be adjusted to the particularities of illness in this patient, taking into account all those things that make this patient an individual – sex, age, occupation, race, social, economic, cultural, and family context. Finally, the information must be shaped at every step by the end, choosing what is good for this patient.

The inherent logical difficulty here is that this patient is something more than an instance of the operation of some scientific principle, although scientific principles are at work in him. Most of the statements we make in clinical medicine are statements of probability applicable to a population of patients sharing certain characteristics with this patient. We can never know in advance where on the frequency distribution curve this patient will be. Yet, to be right, our actions must predict as closely as possible where he is located in the universe from which our probability statements were derived [12].[7]

Optimization of decision and action in the face of uncertainty is the central characteristic of clinical decisions. That is why centuries ago Celsus called medicine an 'art of conjecture'. The degree of uncertainty can be reduced by access to the best available information, by meticulous attention to the clinical arts – history taking, physical examination, by critical use of probabilistic and modal logic and mastery of the art of clinical dialectics. As I have shown elsewhere, the anatomy of a clinical decision is a complicated interplay of several different modes of reasoning and inquiry, not just a scientific one as the biomedical model supposes. Moreover, each step in clinical judgment must be modulated by value desiderata – those that advance the best interests of the one who is ill [18].[8]

These value desiderata are what convert a technically right decision into a good one. A good decision not only cures the illness but does so in harmony with the patient's conception of the conditions of an acceptable life. The choices must be responsive to the way the patient wishes to spend *his* life. This is a moral imperative particularly when the illness is untreatable or fatal. A good decision in this sense is impossible without restoring to the extent possible those human capacities eroded by illness – the capacity to make autonomous choices based on knowledge and free of coercion. An unequivocal criterion of a good decision is the enhancement of the patient's moral agency even when this flies in the face of what science or even a rational bystander might dictate. The patient is the one who must balance his vision of the good life with the realities illness forces upon him.

A good decision must be grounded in some sense on the meaning of illness to this patient. Illness challenges the meaning of life, suffering, and our relationships with others and with God. A patient's whole biography is at stake in any serious illness. Healing is impaired without some attention to that biography. There is a pattern to even the most seemingly chaotic life. The physician must grasp that pattern if his decision is to fit the patient before him. Yet he can do so only partially because the experience of illness is always unique and not penetrable to others.

To be technically right a decision must be objective; to be good it must be compassionate. It is in the fusion of these opposing attitudes that the end of clinical medicine is fulfilled. On the one hand the physician must be able to stand back, analyze, classify, measure and reason. On the other, he must be able to feel something of the experience of illness felt by this patient. He must literally suffer something of the patient's pain along with him for that is what compassion literally means. Often the physician heals himself while healing the patient; oftentimes he cannot heal until he has healed himself. Thus the most ancient of admonitions – physician heal thyself![9]

The medical actions to which right and good clinical decisions lead must be carried out safely, efficiently, and competently. Herein lies the 'art' of medicine strictly speaking. An art consists in the perfection of the execution of a product or art [18].[10] Medicine cannot be defined solely as an art because we can and do delegate specific medical acts to others. It is the decision to undertake a treatment or operation under the conditions of the healing relationship that makes medicine, in Aquinas's sense, a 'ruling art'. Great surgeons are not only good artists but also prudent decision makers about the use of surgery.

IMPLICATIONS OF HEALING AS AN ARCHITECTONIC PRINCIPLE

The significance of any theory of medicine lies in its influence on medical practice, education, and ethics. Perhaps some of the differences in substance between theories can be underscored by an examination of a few of the implications of the healing relationship as an ordering principle in these three areas. I have discussed in part the implications for education elsewhere [20].

Medical Practice

If the distinguishing mark of clinical medicine is the healing relationship then it becomes the central function of all physicians as physicians. Specialists and generalists exercise this function in different ways which must, however, be complementary to each other. Each must answer the four questions that concern any human who is ill: What is wrong? What will happen to me? What can be done? What should be done? The answers to these questions lead to the right and good decision the patient seeks.

The specialist can undertake the healing relationship over only a narrow spectrum defined by the organ, organ system, or technique which bounds his specialty. He is freer than the generalist to define and negotiate the boundaries of his decision making: to help the patient find help in those dimensions of illness outside his specialty.

To the extent that the specialist narrows the boundaries of his field, to that extent he becomes less a physician and more a technician. Also the further his technique takes him from personal encounters the less he is a physician. This is invidious unless the specialist presumes to be a physician in the fuller sense I have defined it here or if he disdains the larger concerns of the generalist his patient may need.

The generalist cannot take refuge in the limitations of his specialty. For him the healing relationship must be entered in the fullest sense. His claim is that he is the 'personal' physician and he must indeed care for the special dimensions of his patient's experience of illness. He must help, care for, comfort and ease when the specialist has nothing to offer. He often starts with a patient whose needs are not yet categorized. He must outline their configuration and devise some plan that places them in some order of priority. The patient often has made the rounds of the specialties; he is still ill, still needing answers to the key clinical questions. Even if the patient's illness has been 'negotiated' out of medicine by other physicians, someone must remain who can help.

The generalist, on this view, is the physician par excellence since he has the most intimate relationship with the healing and helping functions of medicine. A specialist, especially if his domain is a technique, might get away with only scientifically right decisions; but a generalist, never. His most crucial task is to manage the experience of illness, coordinating the recommendations of specialists, interpreting them and, perhaps most important of all, serving as a critic of those recommendations. This critical function flows directly from the need to assist the patient in deciding whether a recommended treatment is worthwhile. It is not often enough performed and obviously a difficult assignment. Yet it is morally unavoidable. The generalist must confine himself to a critical analysis of the value questions in the decision. The technical aspects may be questionable too but they cannot be resolved by the generalist. The opinion of another specialist is needed.

On the view I am proposing, it is clear that contemporary medicine demands a complementary relationship between specialists and generalists. Each has a healing function. As medicine becomes further specialized in the future, the responsibility for the act most specific to medicine – deciding what is right and good for this patient – becomes increasingly the responsibility of the generalist. In this respect Engel's model is more realistic for the generalist, Seldin's for the specialist. Neither suffices unless ordered by the patient's needs for a healing relationship.

Medical Ethics

Once we have decided what medicine *is* and from that what kind of functions physicians perform, we can derive what they ought to do – that is the content of medical ethics. I have suggested a reconstruction of medical morality of the interaction between three facets of the healing relationship – the fact of illness, the act of profession, and the act of medicine [19]. My discussion in this paper has focused only on the third of this triad of relationships – the act of medicine, the right and good healing action for a particular patient.

Obviously the first moral requirement for any ethics of medicine must be competence because without it the physician's promise to help is a lie. Competence is the indispensable requirement for technically right decisions and the first requirement for a good decision as well.

The moral center of medicine, however, occurs at the moment when, having weighed all the things that can be done the physician recommends what *ought* to be done – what is in the best interests of the patient. It is here that value decisions must be made, where the values of physician and patient must intersect and conflict, where their respective moral agencies must be given equal respect. The physician has a moral obligation to conduct the decision reaching process so that the patient's capacities to make his own decision are enhanced to the degree the illness permits. Disclosure of the information needed to make this decision, telling the truth about the patient's condition, doing no harm, keeping promises, helping the patient to make a decision that is free, unmanipulated and authentically his own. Much of this is submerged in the obligation to maintain a morally valid consent and doing no harm. These are all indispensable to the good clinical decision as I have defined it in this essay.

There is an obligation to take the pains necessary to assure that the patient understands the alternatives, the dangers and benefits, the costs, chances of success, limitations of the procedure or treatment. The patient must be helped, too, to make a decision that is authentic, that reflects his values, and he must be helped to discern and identify those values before he decides.

Finally, there is an obligation to help the patient cope with or adjust to a disease that is incurable, progressive, or imminently fatal. The physician need not do all of this himself but he has a major responsibility to see that these dimensions are exposed, and attended to by family, friends, or spiritual advisers. To do this requires some measure of compassion – the capacity to feel something of this patient's experience of illness. Compassion therefore is not just a romantic residuum of ancient medicine, it is a moral imperative of all medicine.

If the healing relationship is the organizing principle of medicine, it imposes moral obligations that go beyond the application of biology to clinical problems. The physician is therefore not entirely free to define medicine as he wishes. Medicine arises in a universal human need that transcends the physician's idiosyncratic definition of its boundaries. It is one thing to determine that a patient does or does not have a medically definable disease – that is the physician's responsibility. It is a different matter to say the patient is not ill. Society supports medicine because it promises to heal and help those who are ill. Society has more than a casual interest in how medicine is defined.

A 'CONCLUDING POSTSCRIPT'

It may be objected, properly, that this essay ignores the responsibilities of medicine beyond those of the clinical encounter. There are, indeed, legitimate functions for medicine such as its social responsibilities, its capacity for biological research, for promoting health and preventing illness, and serving as a source of criticisms of the way we live – individually and collectively. These are socially and culturally important responsibilities for medicine but not necessarily for *clinical* medicine. I have limited myself to what clinicians do, not only for want of space, but also because I wanted to concentrate on what is uniquely the physician's function, one that is unchanging because it is grounded in the universal human need for help and healing of those who are ill. I have addressed some of the other dimensions of medicine elsewhere [25].[11] I recognize that they must be taken into account in any full theory of medicine.

SUMMARY

That theories have practical consequences is no revelation except to the most absolute pragmatists. Equally true is the fact that theoretical confusion lies behind confusion in public and social policy. Nowhere are these truisms more clearly illustrated than in medicine.

Our theories tell us what medicine is, and on this basis we build our models of medical practice and morals. Today we face a vigorous reexamination of all our models of medicine. The consequences for the care of patients and for social and public policy are most significant. Some of the most crucial dilemmas about medicine cannot be resolved without some consensus on what we think medicine is, what it is for, what physicians ought to do and how they ought to do it.

This essay offers a critique of the major contending theories of medicine – agreeing in part with some and disagreeing substantively with others. An architectonics of clinical medicine is offered which is based in the nature of the healing relationship. This is the most permanent, universal and compelling phenomenon of clinical medicine; it gives shape to how medicine should be practiced, what its morals should be, and how its practitioners should be educated.

It is true, as John Burnham has recently emphasized, that the 'Golden Age' of medicine seems to be in decline. The socio-cultural reasons he adduces for this fact are cogent. I would add to them the confusion about theories of medicine which have undergone significant alteration in the same period [3].[12]

Georgetown University,
Washington, D. C.

NOTES

[1] Two journals now deal solely with this subject. *The Journal of Medicine and Philosophy* and *Theoretical Medicine*, both published by D. Reidel, Dordrecht.
[2] See [4]. This recent anthology collects a series of essays defining medicine, illness, health, and disease. The range of definitions past and present indicates how problematic these key concepts remain.
[3] See also [32].
[4] The author likens the proprietary ownership of medical knowledge to the knowledge of baking bread. Michael Bayles [2] says there is no distinction between the physician-patient relationship and that between the car owner and his mechanic.
[5] See also [31].
[6] Given the multiple interpretations in meaning of the terms 'right' and 'good' it is necessary to define my use of them in this essay. Both words are used idiosyncratically, and specifically in relationship to the moment of clinical decision. 'Right' refers to a decision that is technically correct, that squares with empirical and scientific fact, adjusted to the particularities of *this patient*. It is derived objectively by the physician. 'Good' refers to what this patient perceives as worthwhile and valued in his life. It is personally derived. Physicians ordinarily see a 'good' decision as covering both aspects, but patients, in fact, may not. It is to account for both domains – the physician's and the patient's – that the distinction is made. This is not to accept normative relativism but to focus on the moral management of value differences between physician and patient. No argument is made for a particular ontology of 'right' and 'good'. See [7] and [11].
[7] The authors have dilated on medicine as a science of particulars and its inherent fallibility.
[8] See also [26].
[9] This most ancient of maxims can be found in Aeschylus' *Prometheus Bound*, Aesop's *The Worm and the Fox*, the Babylonian Talmud, Cicero, Euripides, the Gospel of Luke – indicating how universally this need is perceived by those to whom physicians deign to minister.
[10] I follow here the classical definition of art as a *recta ratio factibilium*.
[11] See also [22] and [23].
[12] This paper was presented in part as the Second Annual Grant Taylor Lecture, University of Texas Health Science Center, Houston, Texas, April, 1982.

BIBLIOGRAPHY

1. Allen, D. *et al.*: 1980, *Whole Person Medicine*, Inter-Varsity Press, Downers Grove, Illinois.
2. Bayles, M.: 1979, 'Physicians as Body Mechanics', in J. Davis *et al.* (eds.), *Contemporary Issues in Biomedical Ethics*, Humana Press, Clifton, New Jersey.
3. Burnham, J.: 1982, 'American Medicine's Golden Age: What Happened to It?' *Science* **215**, 1474–1479.
4. Caplan, A. J. *et al.* (eds.): 1981, *Concepts of Health and Disease*, Addison Wesley, Reading, Massachussets.

5. Constitution of the World Health Organization: 1958, *Preamble: The First Ten Years of the World Health Organization*, Geneva.
6. de Unamuno y Jugo, M.: 1976, *Selected Works of Miguel de Unamuno*, edited and annotated by A. Kerrigan and M. Nozick, Vol. 6, Novela/Nivola Series LXXXV 6, Princeton University Press, Princeton, New Jersey.
7. Edwards, P. (ed.): 1967, 'The Good', *The Encyclopedia of Philosophy*, Vol. 3, pp. 367–370. Free Press, New York.
8. Engel, G.: 1977, 'The Need for a New Medical Model: A Challenge for Biomedicine', *Science* **196**, 129–136.
9. Engel, G.: 1980, 'The Clinical Application of the Biopsychosocial Model', *The American Journal of Psychiatry* **137** (May), 535–544.
10. Fabrega, H.: 1973, 'Concepts of Disease: Logical Features and Social Implications', *Perspectives in Biology and Medicine* **1** (Summer), 538–617.
11. Frankena, W. K.: 1963, *Ethics*, Prentice-Hall, Englewood Cliffs, New Jersey.
12. Gorovitz, S. and MacIntyre, A.: 1976, 'Toward a Theory of Medical Fallibility', *Journal of Medicine and Philosophy* **1** (March), 51–71.
13. Jacob, F.: 1982, *The Possible and the Actual*, University of Washington Press, Seattle and London.
14. Kass, L.: 1975, 'Regarding the End of Medicine and the Pursuit of Health', *The Public Interest* **40** (Summer), 11–42.
15. Kety, S.: 1974, 'From Rationalization to Reason', *American Journal of Psychiatry* **131**, 957–963.
16. Nozick, R.: 1981, *Philosophical Explanations*, Belknap-Harvard University Press, Cambridge, Massachusetts.
17. Pellegrino, E.: 1976, 'Philosophy and Medicine: Problematic and Potential', *Journal of Medicine and Philosophy* **1** (March), 5–31.
18. Pellegrino, E.: 1979, 'The Anatomy of Clinical Judgments: Some Notes on Right Reason and Right Action', in H. T. Engelhardt, Jr. *et al.* (eds.), *Clinical Judgment: A Critical Appraisal*, D. Reidel, Dordrecht, pp. 169–194.
19. Pellegrino, E.: 1979, 'Toward a Reconstruction of Medical Morality: The Primacy of the Act of Profession and the Fact of Illness', *Journal of Medicine and Philosophy* **4** (March), 32–56.
20. Pellegrino, E.: 1980, *Humanism and the Physician*, University of Tennessee Press, Nashville.
21. Pellegrino, E. and Thomasma, D.: 1980, *A Philosophical Basis of Medical Practice*, Oxford University Press, New York.
22. Pellegrino, E.: 1980, 'The Physician-Patient Relationship in Preventive Medicine: A Reply to Robert Dickman', *Journal of Medicine and Philosophy* **5** (September), 208–212.
23. Pellegrino, E.: 1982, 'The Ethics of Collective Judgments in Medicine and Health Care', Editorial, *Journal of Medicine and Philosophy* **7** (March), 3–10.
24. Pellegrino, E.: forthcoming, 'Biology and Public Administration: Some Touchstones Scrutinized', *Proceedings*, Colloquium Series, C. W. Post Center, Long Island University, Greenvale, New York.
25. Pellegrino, E.: forthcoming, 'Optimizing the Uses of Medical Knowledge', Academy of Independent Scholars Symposium.
26. Pellegrino, E.: forthcoming, 'Value Desiderata in Logical Structuring of Clinical

Diagnosis', in J. L. Peset *et al.* (eds.), *Ethics of Diagnosis*, D. Reidel, Dordrecht.
27. Sade, R.: 1971, 'Medical Care as a Right: A Refutation', *New England Journal of Medicine* 285 (December 2), 1288–1292.
28. Seldin, D.: 1977, 'The Medical Model: Biomedical Science as the Basis for Medicine', *Beyond Tomorrow*, The Rockefeller University Press, New York.
29. Seldin, D.: 1981, 'The Boundaries of Medicine', Presidential Address, *Transactions of the Association of American Physicians* 94, 75–86.
30. Siegler, M.: 1981, 'The Doctor-Patient Encounter and Its Relationship to Health and Disease', in A. Caplan *et al.* (eds.), *Concepts of Health and Disease*, Addison Wesley, Reading, Massachussets, pp. 627–644.
31. Spicker, S. (ed.): 1970, *The Philosophy of the Body*, Quadrangle Books, Chicago.
32. Westberg, G.: 1979, *Theological Roots of Wholistic Health Care*, Wholistic Health Centers, Inc., Hinsdale, Illinois.
33. Whitbeck, C.: 1981, 'A Theory of Health', in A. Caplan *et al.* (eds.), *Concepts of Health and Disease*, Addison Wesley, Reading, Massachussets, pp. 611–626.

JOSEPH MARGOLIS

THE PSYCHIATRIC PATIENT-PHYSICIAN RELATIONSHIP

It is generally appreciated by those who examine social institutions carefully that the function of legal punishment is quite obscure: it actually seems to contribute to dramatic forms of recidivism; it has a rather poor record of rehabilitation; it seems to have almost no significant deterrent force; it may aggravate the tendency toward even more serious crime; and it is characteristically managed in a brutal, unproductive, or mindless way. No one would claim that psychiatric treatment deserves to be similarly criticized, although Thomas Szasz of course, quite some time ago, brought charges against the profession that were at least well on their way toward blander versions of a similar complaint [26].[1] And the worries about the political use of psychiatry, notably focused by the case of Zhores Medvedev, in the Soviet Union,[2] oblige us to speak more directly of the function of psychiatry – hence, of reasonable constraints and directives regarding the patient-physician relationship. Of course, no one would deny that provision for a system of institutionalized punishment is conceptually entailed by a system of criminal law; correspondingly, no one would deny that, conceding the reality of psychiatric needs and disorders, provision for a system of institutionalized care and delivery is entailed by the life of a society committed to the individual and aggregated care and well-being of its members. We need not labor the analogy further.

Nevertheless, the conceptual uncertainties surrounding the professional practice of psychiatry are pretty obvious. If confirmation of these uncertainties were needed, we could hardly collect more convincing evidence than that inadvertently provided by the opening essay of that most successful volume, *The Harvard Guide to Modern Psychiatry*. The editor-in-chief, Arman Nicholi, specifically discussing the therapist-patient relationship in a book billed as focused on the patient, says without qualification:

> Though he may express the same amenities in both his professional and social relationships, the skilled clinician seldom confuses one with the other. He realizes that as long as his patient remains his patient, the therapist-patient interaction must, for the benefit of the patient, remain within the context of a professional relationship; this relationship differs from the social one in that it exists to perform a unique task – namely, to evaluate and treat the conflicts prompting the patient to seek psychiatric help. . . . Whereas a

Earl E. Shelp (ed.), The Clinical Encounter, 173–186.

social relationship focuses on the needs of both parties, a professional relationship focuses solely on the needs of the patient – never on the needs of the therapist ([23], p. 6).

Nicholi favors the notion of a 'therapeutic alliance' or 'therapeutic contract'. He specifically cites the Hippocratic precepts, and his own view is clearly Hippocratic; but he never raises the standard objections to the Hippocratic tradition.

Now, the general counterargument against Hippocratism is quite straightforward: *if* the physician-patient relationship – *a fortiori*, the psychiatrist-patient relationship – is a professional one (as Nicholi stresses and most would concede), then the code of conduct and proper function of psychiatrists *cannot but reflect an entire society's reasoned view of the service that profession is to provide*. But then, even if it is supposed that the sole function of the physician is to 'benefit the sick' (as the Hippocratic Oath has it), the principle is, first, derived and authorized from a larger notion of social well-being and, secondly, is always in principle open to enlargement and revision in the name of that same interest. Robert Veatch quite reasonably argues against the Hippocratic principle on the grounds that it is 'consequentialistic', 'paternalistic', and 'individualistic' ([27], Chs. 1, 6).[3] Even the priorities of patient care cannot be decided on Hippocratic grounds, which effectively indicates the incompleteness and distortion of Nicholi's formula. The same may be said of the social price of patient care.

It is revealing that Nicholi warns us only regarding the intrusion of the therapist's needs: he somehow fails to consider that the needs of particular social groups (families, for instance) or of a larger society may have legitimate claims on the professional function of psychiatry and medicine in general. Furthermore, Nicholi holds that the 'unique task' of the psychiatrist is "to evaluate and treat the conflicts prompting the patient to seek psychiatric help." This complicates matters unnecessarily: for, for one thing, the term 'conflict' obscures the difference between psychiatric diseases and complaints that are psychiatrically manageable but are not diseases; and for another, it fails to distinguish between the initiative of voluntary patients and of other responsible and interested parties. In short, the conceptual puzzles regarding the therapeutic relationship in psychiatry involve both the social responsibility of medical professionals adjusted to psychiatry and the distinctive nature of the range of complaints that fall within the practice of psychiatrists. But although the two issues are quite distinct, they are effectively inseparable.

Psychiatry may well be the single large professional undertaking in which services rendered are expected to be justified as canonical or objective forms of treatment of complaints that, by their very nature, draw physician and patient into intimate emotionally charged relationships threatening (and calling into question) the very possibility of an objective professional practice. Something similar occurs in the law and in general medicine. But in the law, the initiating complaint usually involves a third party, and professional practice is in some measure at least directed to exploiting, within the boundaries of acceptable practice, the partisan advantage of one's client. In general medicine, the initiating complaint is primarily formulated in somatic terms, and psychological complications between physician and patient tend to be viewed rather more as intrusions. But in the psychiatric relationship, normally, the trick is to be able to preserve, within transactions inevitably marked by significant transference and countertransference, canonical practices that may claim a measure of objectivity favorably approaching that of somatic medicine. Doubts arise precisely because of the intensity of the therapeutic relationship focused on complaints that, for conceptual reasons, are either not fixed with sufficient certainty as medical disorders or are not explicitly matched with suitably canonical plans of treatment or both. This is particularly marked where psychogenic causes are thought to be decisive, where the principal complaints concern thought processes and cognate behavior, and where the identified syndromes converge with behavior viewed independently as deviant, or psychologically or socially threatening or unacceptable, on grounds in accord with prevailing taste, moral and religious conviction, traditional practice, political tolerance, and the like. There is, in short, a very strong ideological or ideology-like dimension in psychiatry that cannot fail to affect any reasonable overview of the therapeutic relationship and of how it ought to be managed. This feature appears to be quite ineliminable.

There is one global proposal for governing therapeutic practice that recommends itself, prima facie, as fairminded – particularly, conceding the partisan possibilities of psychiatry. That is the principle that: (1) insofar as professionals are committed to serving the medical needs of their patients, the norms of psychiatric health should conform with medical (that is, psychiatric) consensus; (2) that where they are contested and tend to resemble the norms of acceptable behavior and well-being in society at large, such norms should conform reasonably closely with the prevailing values of the society; and (3) that, within this latter range and also where psychiatric service is less specifically medical in nature though still distinctly professional, service should conform as closely as possible to the values and interests of clients or of those who are effectively their guardians or have responsibility for their well-being.

This proposal is so global, however, that it is well-nigh incontestable without really providing for the resolution of familiar quarrels. Furthermore, though it is formulated as a universal principle, its clauses are such that, in effect, it obscures the strong possibility that the rationale underlying psychiatric canons is both pluralistic and relativistic: pluralistic, in the sense that the requisite canons may have to accommodate divergent traditions within complex modern societies; relativistic, in the sense that the principle implicitly concedes that there is no straightforward way in which to determine uniquely valid norms of psychiatric health – if not for the entire range of, then at least for a significant range within, the standard practice of psychiatry. Here, we have broached an entire nest of puzzles. Still, informally, one might offer as instances of the kinds of cases to be subsumed under (i)–(iii), the following – drawn from the *DSM–III* [1] : (i) such organic brain syndromes as dementia, notably those associated with Alzheimer's and Pick's diseases; (ii) such so-called substance use disorders as alcohol dependence and abuse, or similar patterns involving opioid, sedative, symptomimetic, and hallucinogenic dependence and abuse; (iii) so-called ego-dystonic homosexuality.[4]

It is reasonably clear that professional psychiatry is disposed to avoid the appearance of merely accommodating contingently pluralistic patterns of behavior or, worse, conceding the relativistic nature of psychiatric medicine. To some extent, a rather strong argument can be mounted for the ineliminably ideological (and therefore partisan) dimension of even somatic medicine (or at least of a significant part of it), without disturbing the core consensus on which it rests.[5] Put most briefly, medicine is a normative discipline, insofar as it is committed to diagnosing and treating patients in accord with what it takes to be the functional norms of the human person. Those cannot fail to accommodate the fact that the concept of a person is not the same as the concept of a recognizable specimen of *Homo sapiens*; and that, in particular, whatever the proper norms of the functioning of persons may be, they cannot be identical with those of the functioning of a mere biological (culturally undeveloped) specimen of *Homo sapiens.*[6] It is also clear, however, that, given some reasonably straightforward picture of the needs of higher animal life and species-specific functioning, a fairly stable range of forms of somatic functioning is bound to be supported by most competing models of somatic medicine – *a fortiori*, of psychiatry. Nevertheless, unless there is a strong cognitive procedure for discovering the proper norms of medicine, that favorably compares with procedures in nonnormative empirical science (which seems untenable[7]), or unless a compelling case can be made for the nonnormative nature of medicine itself, including psychiatry,[8] it seems impossible to eliminate tendencies in accord with pluralism and relativism (which, of course, are not at all the same). One reason for this is simply that the norms of *species-specific* health among humans must be extrapolated from the highly variable and technologically plastic forms of cultural life that human persons have developed and could develop.

DSM–III, which is theoretically bolder than *DSM–II*, is noticeably cautious about the notion of a mental disorder and actually fails to supply an account of 'dysfunction', on which 'mental disorder' itself depends. In its Glossary of Technical Terms, *DSM–III* offers the following:

> In *DSM III*, a mental disorder is conceptualized as a clinically significant behavioral or psychological syndrome or pattern that occurs in an individual and that typically is associated with either a painful symptom (distress) or impairment in one or more important areas of functioning (disability). In addition, there is an inference that there is a behavioral, psychologic, or biologic dysfunction, and that the disturbance is not only in the relationship between the individual and society. When the disturbance is limited to a conflict between an individual and society, this may represent social deviance, which may or may not be commendable, but is not by itself a mental disorder ([1], p. 363).

There are of course quite a number of *caveats* here. Construed as a definition, this entry has been criticized in the most sustained and interesting way by Charles Culver and Bernard Gert. Employing 'malady' as a term more inclusive than disease, illness, disorder, dysfunction, and the like, Culver and Gert offer their own improved definition:

> A person has a malady if and only if he has a condition, other than his rational beliefs and desires, such that he is suffering, or at increased risk of suffering, an evil (death, pain, disability, loss of freedom or opportunity, or loss of pleasure) in the absence of a distinct sustaining cause ([7], p. 81).

They clearly regard their own definition as 'universal and objective' and hold that the values integral to the concept of a malady are also 'universal' ([7]. p. 81). They are somewhat too strenuously committed to essentialist or universalist definitions – which many have reasonably argued are not required or even (relative to diachronically developing disciplines) plausibly to be pursued.[9] But they also correctly note weaknesses in the *DSM–III's* definition – and they themselves attempt to supply a notion of mental malady that falls within their own larger notion. Nevertheless, their definition is noticeably indecisive at precisely the point at which (most reasonably) they criticize the *DSM–III*. In fact, there is a telltale slippage in their account, which in effect *no* current analysis has managed to overcome successfully (that is, with an eye to avoiding the threats of relativism – consequently, dangers to essentialist definitions). The point is of some importance, since it bears directly on the prospects of formulating objective constraints on the therapeutic relationship.

First of all, Culver and Gert correctly notice the difficulty of specifying psychologic or behavioral dysfunction as opposed to biologic dysfunction. The latter, they say, "is reasonably clear when applied to organs of the body, such as the liver, kidney, or heart" ([7], p. 94). The dysfunction of *persons* seems (to them) to be not derivable from, or restricted to, the dysfunction of biological organ systems. Fair enough. Secondly, they themselves (with regard to their own definition) simplify matters plausibly by insisting (*contra* the *DSM–III*) that "there seems to be no useful distinction between a disability and a dysfunction: the disability is the dysfunction" ([7], p. 94). They also insist that "there is no fundamental difference in kind between physical and mental maladies" ([7], p. 91); that is, each exhibit the essential defining traits of maladies, which is to say, effectively, that malady is a state of persons, not of organs. But when they come to explain what a disability (or dysfunction) is, they simply say that "a person is suffering a disability when he lacks an ability that is characteristic of the *species*, or when he has an extraordinarily low degree of that ability" ([7], p. 77; italics added). For example, if "an overwhelming majority of nondiseased, noninjured members of the species . . . have had [the] ability [to walk ten miles, then it is] characteristic of the species [and its lack signifies a disability]" ([7], p. 77).[10] Culver and Gert fail to notice that: (a) even if there are such abilities characteristic of the members of *Homo sapiens* as such, it does not follow that there are comparably clear abilities characteristic of all *human persons* or are such adequate to the notion of medically relevant malady or disability; (b) even if, in accord with (a), therefore, there is a conservative core of somatic medicine that hardly varies from one society to another, their definition obscures the fact that, particularly with respect to mental and behavioral disabilities (psychiatrically pertinent disabilities), their own thesis is *not* (or not likely to be) universal in scope, but actually adjusted to the plural, ideologized functions differentially favored in one society or another; (c) they do not themselves provide a reliable clue to the functions of culturally developed persons as distinct from the behavioral or holistic functions of mere members of a biological species; and (d) there *is* no nontendentious way in which to specify the functions of *persons* as opposed to the functions of organs or of the aggregated members of *Homo sapiens*. Longevity and level of biological function cannot be freed from technological contingencies; malady cannot convincingly be restricted to problems of longevity and reproduction; *and* any attempt to do so, either in Culver and Gert's manner or, more drastically, in Christopher Boorse's, must contend with the charge that the selection of statistical patterns is itself a form of normative preference *as well as* one that inevitably depends on culturally favored values not reducible to the species-specific. There is nothing suspect in this; but it confirms the sense in which psychiatry – hence, the norms of therapeutic practice – cannot be freed from complications due to partisan, ideologically skewed, and relativized values. That the resultant system (psychiatry) is moderately relativistic neither entails its incoherence nor threatens its objectivity (in a sense).[11] These considerations help to clarify the possible vacuity of the global proposal (i)–(iii) offered earlier on.

The most compelling evidence regarding the partisan and tendentious nature of psychiatry – particularly as one approaches cases falling within (iii) – is of course supplied by the history of the American Psychiatric Association's classification of sexual disorders, particularly homosexuality. The entire drama of deleting homosexuality as a medical disorder in *DSM–II* and its rather anomalous position in *DSM–III* (where, for instance, "ego-dystonic heterosexuality" is not acknowledged) is by this time well known.[12] Here, the impossibility of fixing a strongly convergent model of mental and behavioral health – or freedom from malady – ranging over the usual scope of psychiatric practice is reasonably obvious. It is also worth remarking that suicide is not mentioned in *DSM–III*, though there can hardly be any question that professional psychiatry – particularly, psychoanalytically oriented psychiatry – has been strongly disposed, if not positively committed to viewing suicide attempts as evidence of an antecedent psychotic break or at least as plainly disabled with respect to rational powers.[13]

Culver and Gert are not clear on the matter. Their intuition favors the reasonableness (and rationality) of suicide under certain conditions. But their definition of malady makes death one of the essential evils (which is surely anomalous): "all persons [they say] avoid death, pain, and disability *unless they have an adequate reason not to avoid them*" ([7], p. 70). Their remark risks vacuity; but one can see part of their point if (as they suggest) one might "seek death when life has become too painful" ([7], p. 70). Still, they are not clear about cases that do not involve intolerable pain. Where, for instance, they have disabilities in mind, they characteristically fail to distinguish between those proper to *Homo sapiens* and those proper to human persons. One must remember that their account of malady is given in terms of 'evils'. They actually say that "The seriousness of an evil is determined by several factors, among which is the impact it will make on the person's life at a given time. On this test, death is clearly the most serious evil, for nothing has greater impact than death" ([7], p. 169). This would surely lead one to characterize suicide as a malady, except perhaps in those cases (far from exhausting the usual range) in which pain is intolerable or one's disabilities are well below the level at which one could expect to function as a person (Karen Ann Quinlan's case, perhaps). They fail to come to grips with the issue, though they consider suicide.

The matter is quite important, because it draws us on to the second source of complication (noted earlier) pertinent to assessing the psychiatric transaction – the social responsibility and the social and individual services of the professional psychiatrist. Culver and Gert actually use their notion of evils – chiefly, death, pain, and disability – in formulating their well-known defense of medical paternalism ([7], Chs. 7–8). It is a *necessary* condition of paternalism, on their view, that the paternalistic agent *benefit* the patient or subject by means of his action ([7], p. 130). The objectivity of the (prudentially specified) evils they mention in their definition of malady and the rational ordering of those evils provide (on their view) a conceptually satisfactory basis for paternalistic intervention – that is, intervention against or without the voluntary consent of the subject or his guardian or the like. But then, either suicide (on the argument) should be prevented because it is a malady; or (like homosexuality) its classification merely justifies the rationalized intrusion by a professional body or by society at large into the private lives of particular persons. The distance between such interventions and the political use of psychiatry is not at all easy to establish; and competing views of prudential 'evils' (perhaps adjusted culturally, as seems predictable) would justify competing views both of paternalistic intervention *and* of the primary social and individual services 'correctly' rendered by the psychiatrist in different societies. In any case, what we see here is a transparent use of culturally partisan values treated as if they were genuinely universal *at the level of personal life*.

Once we see the sort of slippage possible, it takes very little effort to see how cases belonging to category (ii) – even to (i) – may succumb to paternalistic and related arguments. For, not only are the norms of personal-level functioning not clearly definable in the universalized form favored in psychiatry; the norms of personal-level evils are also not definable in any determinate way that could support objective and universalized appraisals and rankings. Even beyond that, once we consider (as Culver and Gert insist) 'increased risk' of malady (or attendant evil), we are bound by the logic of the argument to construe clearly debatable cases as bona fide cases of malady, just because of the incipiency of their apparent symptoms – hence, as 'treatable' in a way that invites all the dangers here sketched. Nevertheless, the consensus with which *extreme* cases within (i)–(iii) invite a judgment of mental illness or malady *does not as such confirm the essential properties of malady or such malady, and therefore does not permit us to claim an objective basis for identifying malady in (seemingly) incipient cases*. For example, incipient but controlled alcoholic dependence is *not*, for that reason, an incipient form of mental malady, even if a flaming case of alcoholism is fairly regarded as a form of malady. In the same spirit, the authors of *DSM–III* note the omission of the *DMS–II* diagnosis category of the neuroses, observing only that "At the present time . . . there is no consensus in our field as to how to define 'neurosis'. . . . The term *neurotic process* should be used only descriptively" ([1], p. 9). Somewhat the same observation *might* have been made of the term 'psychotic', which the authors define as "indicating gross impairment in reality testing" ([1], p. 367). It is difficult to see how this can escape ideological skewing altogether, in spite of the fact that there is little doubt that certain extreme cases both of psychosis and neurosis are clinically beyond dispute.

It may well be that the concept of a psychiatric disorder, disease, or malady is systematically different from that of a somatic disorder – but not (as in Szasz's sense) such as to call into question the legitimacy of the very concept. Category (iii) cases may not be maladies at all (it is true), though they may be precipitating factors or the manifestation of malady. Characteristically, such cases entail presumptions about the correct or essential functioning of human persons as such -- for instance, as in medically motivated suspicions about homosexuality, suicide, marked deviance along behavioral, affective, social, interpersonal, and cognitive lines, particularly as viewed in accord with so-called V Codes ([1], pp. 331–334). Here, there is a distinct convergence between medical and moral-like values bearing directly on norms of personal development and self-realization. Category (ii) cases may be admissible as psychiatric disorders. But they are logically distinctive in that: (1) *only* marked or extreme cases are maladies – so that incipient or mild manifestations are *not* (as such) 'early' or 'benign' phases of such disorders (contrary, say, to the sense in which early stages of cancers are benign phases of distinct maladies); and (2) such maladies reflect in (at least) a moderately relativistic way the developmental values favored and integrated in one culture or another. Category (i) cases are similar to standard cases in somatic medicine; indeed, they can be assigned an organic etiology of a predisposing or precipitating nature. But even here, particularly where the primary evidence is behavioral rather than organic, where the onset is mild or incipient, or where clinical manifestations strongly resemble those of category (ii) or the V Codes (the dementias, for instance, so-called 'substance-induced organic mental disorders' like alcohol intoxication, amnesia, delusion, affective disorders associated with manic and depressive episodes, even psychotic episodes), the line of demarcation between (i) and (ii) is most difficult to draw. There is often so much uncertainty here that there is good reason to be wary of medically motivated interventions where objections appropriately addressed to category (ii) and (iii) cases seem warranted.

In fact, in the entire range of psychiatric cases, with rather few (but not negligible) kinds of exception, *there is no secure basis for distinguishing medical and moral (or legal or political) intervention*. This is complicated, further, by the compelling fact that there is no known conceptual basis (short of social consensus and frankly doctrinal and partisan conviction) on which to ground a psychiatrically adequate model of personal development and well-being; and neither consensus or doctrinal conviction can insure objectivity, universality, the avoidance of relativism, rationalized but otherwise indefensible or partisan intervention, or the like.

The bearing of all this on the defense of psychiatric paternalism is obvious. Culver and Gert concede no difficulty, since, on their view, the psychiatrist (or physician in general) must benefit his patient; but on their theory of objective 'evils' and the objectivity with which evils can be ranked (within limits), intervention can be straightforwardly defended. The issue is a vexed one, because of the disputed role of voluntary behavior or capacity as a constraint on paternalism. Tom Beauchamp, for instance, rejects all forms of paternalistic intervention [2],[14] but he does so partly because he does *not* oppose the sort of intervention Culver and Gert support; *he* supports such cases because he does not regard them as involving paternalistic interventions, since (on his view but not theirs), the agent does not exhibit the requisite capacity for voluntary behavior. On his view as on theirs, the objectivity with which harm or threatened harm can be detected is still reasonably clear. Culver and Gert are sanguine because they are also convinced that the voluntary is conceptually straightforward. They dispute Beauchamp's contention that the central cases do not involve the voluntary; and they correctly draw attention to the important issue that a further puzzle remains about "when and when not to interfere with a person's nonvoluntary action to harm himself" ([7], p. 145). But the challenge to a ready acceptance, on medical grounds, of psychiatric initiatives within the context of professional practice – in particular, of paternalism – rests with: (a) the difficulty of separating clinically specified harm and harm defined in terms of moral, legal, political, religious, and similar concerns; (b) pluralistic and relativistic grounds on which, beyond the most generic prudential concerns (pain, death, and disability, for instance), determinate harm can actually be objectively specified; (c) conceptual uncertainties of a profound sort affecting the very notion of the essential functions of human persons as such; (d) the conceptual complexity of the notion of voluntary behavior and its criteria;[15] (e) the increasing complication of cases ranging through categories (i)–(iii); (f) reasonable presumptions, under these circumstances, in favor of voluntary decisions of patients or of their personal interests and preferences at least; and (g) the extremely untidy matter of the relationship between the interests of patients, certain relatively intimate groups affected by their lives (families, for instance), and the society at large *and* between the psychiatrist's obligations with respect to the primary patient, such intermediate groups, and the society at large.

Finally, regarding (g), there is and can be no clear conceptual procedure for formulating an objective canon. There can be no conceptual strategy for deciding in a strongly convergent way even the most salient conflicts between an individual patient's, his family's (or other intermediate group's), and the society's 'interests' or ranking of interests; at least there cannot be such a strategy justified independently of the doctrinal and partisan convictions internal to the historical life of the very society in which such problems arise. It *is* already an enlightened discovery to have recognized the complexity of the matter – the fact, that is, that an exclusively Hippocratic solution (favored, for instance, by Nicholi) is simply one extreme, quite partisan alternative. And any global defense of the moral or related values by which all societies ought to be guided cannot fail to address the plain record of the disputatiousness of every such effort.[16]

More narrowly construed, one can see that at least a strong prima facie argument can be made for maximizing confidentiality and compliance with a client's choices or interests in cases, within categories (i)–(iii), that favor – in the respects already sketched – features most prominent in (iii); that, with cases centered in category (ii), the custodial, emotional, economic, and similar interests of affected intermediate social groups (the immediate family, for instance) must be accorded due consideration, potentially affecting confidentiality and compliance; and that, for the entire range of psychiatric cases but tending particularly toward the pole of category (i) and pronounced instances of (ii), the overriding interests of society at large – political, legal, (frankly) doctrinal and ideological, certainly economic –

cannot be convincingly ignored or unconditionally subordinated. For example, as Veatch reports, the American Medical Association has, since 1957 and through the latest (1980) revisions of its principles, recognized that "the responsibilities of the physician extend not only to the individual, but also to society, where these responsibilities deserve his interest and participation in activities which have the purpose of improving both the health and the well-being of the individual and the community." This is taken by the AMA itself to extend to breaking a patient confidence in order to protect the welfare of the community ([26], pp. 158–159). Clearly, in cases involving involuntary commitment and statutory obligations on psychiatrists not to resist court-ordered disclosures of otherwise confidential material, the presumptive interests of the state and of civil society are explicitly recognized.[17] Extreme arbitrariness, various forms of opportunism, inconsistency of principle, and undue narrowness of conception are undoubtedly detectable even here – particularly, within social systems but also to some extent across systems. Nevertheless, it seems quite impossible to fix, by principled criteria, the correct priorities regarding the interests and claims of individual patients and clients, intermediate social groups affected, professional psychiatry, and the state and society at large – that is, in any way neutral to, and objectively independent of, the ideological orientation of the society in question. At the very least, every society must concern itself with the primary use and funding of such a large service profession as psychiatry relative to the historical circumstances in which it finds itself. For example, one can imagine that the widespread occurrence of depression, anxiety, and related disorders, perhaps occasioned by serious economic hardship, war or incipient war, natural disaster or the like, might well, in accord with a prevailing ideology, be judged to require a substantial reordering of the usual priorities and the usual management of priorities respecting the accessibility of psychiatric care. These are matters that the Hippocratic precepts largely ignore (at least as they have been construed in modern times).

This, then, is perhaps as much as one can say, without betraying a partisan commitment of one's own. No one, of course, can really escape such a commitment; but it seems more useful, here, to clarify the conditions under which competing policies may be rightly understood than to plead one's own conviction.

Temple University,
Philadelphia, Pennsylvania

NOTES

[1] Cf. [9].
[2] See [21].
[3] Cf. also [8].
[4] See [1].
[5] See [15].
[6] See ([17], Chs. 1, 6).
[7] Cf. [11]. The argument can be extended without difficulty to all cognitive claims regarding normative values.
[8] Perhaps the strongest case for the nonnormative nature of medicine (and psychiatry) is advanced by Christopher Boorse [4] and [5]. Boorse also promises a forthcoming book, *The Concept of Health*. I find Boorse's argument uncompelling, however; see [19].
[9] Cf. [23].
[10] But see p. 183: "*Person* is not a biological concept. . . ."
[11] I have developed the concept of relativism for judgments in the arts; see [19]. And I have generalized the account in an as yet unpublished paper, 'The Nature and Strategies of Relativism'.
[12] See [12], [14], and [20].
[13] Cf. [14].
[14] Cf. [3] and [6].
[15] It is worth noting, for instance, that when they attempt to define a 'volitional ability' (presumably the ability to act as a voluntary agent), Culver and Gert introduce the undefined concept of "the volitional ability to will to do X" ([7], pp. 109–113). This poses a serious difficulty for Culver and Gert, because of their insistence on clear essential definitions. Beauchamp and Childress are more casual about defining the voluntary – and for that reason more plausible – but what they say strengthens our sense of the difficulty of sharp criteria; cf. [3], also ([9], Ch. 4).
[16] Cf. [10] and [18].
[17] Cf. [24] and ([25], Ch. 31).

BIBLIOGRAPHY

1. American Psychiatric Association: 1980, *Diagnostic and Statistical Manual of Mental Disorders*, 3rd ed. [DSM–III], Washington, D. C.
2. Beauchamp, T. L.: 1977, 'Paternalism and Bio-Behavior Control', *The Monist* **LX**, 60–80.
3. Beauchamp, T. L. and Childress, J. F.: 1979, *Principles of Biomedical Ethics*, Oxford University Press, New York.
4. Boorse, C.: 1975, 'On the Distinction Between Disease and Illness', *Philosophy and Public Affairs* 5, 49–69.
5. Boorse, C.: 1976, 'What a Theory of Mental Health Should Be', *Journal for the Theory of Social Behavior* **VI**, 61–84.
6. Childress, J. F.: 1979, 'Paternalism and Health Care', in W. L. Robinson and M. S. Pritchard (eds.), *Medical Responsibility*, Humana, Clifton, New Jersey, pp. 15–27.

7. Culver, C. M. and Gert, B.: 1982, *Philosophy in Medicine: Conceptual and Ethical Issues in Medicine and Psychiatry*, Oxford University Press, New York.
8. Edelstein, L.: 1967, *Ancient Medicine: Selected Papers of Ludwig Edelstein*, O. Temkin and C. L. Temkin (eds.), John Hopkins University Press, Baltimore.
9. Margolis, J.: 1966, *Psychotherapy and Morality*, Random House, New York.
10. Margolis, J.: 1971, *Values and Conduct*, Clarendon, Oxford.
11. Margolis, J.: 1975, 'Moral Cognitivism', *Ethics* **LXXXV**, 136–141.
12. Margolis, J.: 1975, *Negativities: The Limits of Life*, Charles Merrill, Columbus.
13. Margolis, J.: 1975, 'The Question of Homosexuality', in R. Baker and F. Elliston (eds.), *Philosophy and Sex*, Prometheus Books, Buffalo, pp. 288–302.
14. Margolis, J.: 1976, 'The Concept of Disease', *Journal of Medicine and Philosophy* **1** (September), 238–255.
15. Margolis, J.: 1976, 'Robust Relativism', *Journal of Aesthetics and Art Criticism* **30**, 37–46.
16. Margolis, J.: 1978, *Persons and Minds*, D. Reidel, Dordrecht.
17. Margolis, J.: 1979, 'The Prospects of an Objective Morality', *Social Research* **XLVI**, 744–765.
18. Margolis, J.: 1980, 'The Concept of Mental Illness: A Philosophical Examination', in B. A. Brody and H. T. Engelhardt, Jr. (eds.), *Mental Illness: Law and Public Policy*, D. Reidel, Dordrecht, pp. 3–23.
19. Margolis, J.: 1982, 'Homosexuality', in T. Regan and D. VanDeVeer (eds.), *And Justice for All: New Introductory Essays in Ethics and Public Policy*, Rowland and Littlefield, Totowa, New Jersey, pp. 42–63.
20. Medvedev, S. and Medvedev, R.: 1971, *A Question of Madness*, E. de Kadt (trans.), Alfred A. Knopf, New York.
21. Nicholi, A. M., Jr.: 1978, 'The Therapist-Patient Relationship', in A. M. Nicholi, Jr. (ed.), *The Harvard Guide to Modern Psychiatry*, Harvard University Press, Cambridge, pp. 3–22.
22. Putnam, H.: 1975, *Philosophical Papers*, Vol. 2, Cambridge University Press, Cambridge.
23. Stone, A. A.: 1975, *Mental Health and Law: A System in Transition*, Government Printing Office, Washington, D.C. – DHEW Publication No. ADM 75–176.
24. Stone, A. A.: 1978, 'Psychiatry and the Law', in A. M. Nicholi (ed.), *The Harvard Guide to Modern Psychiatry*, Harvard University Press, Cambridge, pp. 651–664.
25. Szasz, T. S.: 1961, *The Myth of Mental Illness*, Hoeber-Harper, New York.
26. Veatch, R. M.: 1981, *A Theory of Medical Ethics*, Basic Books, New York.

ROBERT M. VEATCH

THE PHYSICIAN AS STRANGER: THE ETHICS OF THE ANONYMOUS PATIENT-PHYSICIAN RELATIONSHIP

In virtually everything that is written, said, or thought about the patient-physician relationship the ideal of a close, long-term, concerned, all-knowing commitment is lurking. It is recognized that not all relationships between health professionals and lay people live up to that ideal. In fact, all too often physicians are condemned for being distant, aloof, unconcerned about the life the patient leads in all its richness of beliefs, values, lifestyle, social involvements, and psychological problems. The 'physician as friend', however, has been the ideal, and much of our medical ethical reflection, both formal and folk reflection, has been oriented toward that ideal.

In contrast, a great deal of health care is delivered in a model in which the physician is a stranger. This is not only, not primarily, because physicians are not warm, friendly, caring beings. Everything we can observe about them seems to indicate that their capacity for warmth and friendship is probably no better nor worse than any other sociological group. Rather the institutional structure of the health system available to increasing numbers of people today dictates that health care will often be delivered between strangers. Care is given in inner-city clinics, rural health centers, student health services, military and veterans' hospitals, tertiary care centers, and in the offices of specialists seen for highly technical one-time referral consultations.

No one today is exploring what difference it makes in the medical sociology and the medical ethics of the patient-physician relationship whether the physician is assumed to be a friend, a long-term acquaintance knowledgeable about the beliefs, values, and lifestyle of the patient, or whether the institutional constraints on the delivery of health care mandate that the physician will necessarily be a stranger. It may well be that, regardless of the medical ethical tradition in which one places oneself, it ought to make a difference whether the physician and the patient relate to one another as friends or as strangers. The nature of those medical ethical differences will be the focus of this essay. The primary objective, therefore, is to explore the differences it should make in medical ethical theory whether the 'stranger' or the 'friend' model is operative, not to argue for the moral superiority of one or the other. Only in the final section of the essay will the normative dimension be broached, where it will be asked whether the traditional preference for the

Earl E. Shelp (ed.), The Clinical Encounter, 187–207.

physician as friend is warranted or a case can be made for a morally normative distance between medical professional and lay person.

THE FRIEND/PHYSICIAN AS THE IDEAL

By labelling one model of the ideal patient-physician relationship as the model of the 'friend/physician' I have certain characteristics of friendship in mind. In the terms of Talcott Parsons ([31], pp. 171–177) the friendship relationship is functionally diffuse. It involves knowledge and even responsibility across a wide range of issues and many spheres of the individual's life. It contrasts with functional specificity, in which the physician and patient would know each other and interact only in the narrower sphere of medicine. The character traits of the relationship would be those of traditional friendship: compassion, kindness, sympathy, warmth, and fidelity. These stand in contrast to the character traits stereotypically attributed to the modern, more specialized, businesslike physician: efficiency, technical competence, impartiality, coldness, and distance.

The Friendship Metaphor

I am using the imagery of friendship somewhat as a metaphor recognizing that certain characteristics of friendship do not really apply. Notions of status equality, for example, normally would not apply. The friendship of the ideal physician/patient relationship has more distance to it, more objectivity, than the more traditional friendship. The physician shows a concern that has been described as 'detached' [17]. While the physician is concerned about the plight of the patient, he or she ought to remain sufficiently detached that emotional involvement cannot cloud the objectivity needed for good medical practice. Social interaction across a wide range of institutions outside of medicine such as the church, schools, social clubs, and the like would be considered optional, but perhaps a desirable extra. Many essential characteristics of friendship, however, such as a long-term ongoing relationship, knowledge of family, and sensitivity to special interests are included.

There is little doubt that the ideal physician is often considered a friend in this special, qualified sense in which I am using the term. Political scientist Richard Flathman, for example, says of physicians, "if they are *good* physicians, they have a close, perhaps even an intimate, knowledge of their patients – knowledge extending well beyond their medical histories" ([16], p. 108). Physician Manfred Bleuler writing in a journal with the remarkably unfriendly title *Diseases of the Nervous System* makes explicit use of the friendship metaphor: "In the patient's pain, in his despair, in his misery, the patient called for a friend, a friend whom he can trust, a friend whose wisdom, whose willingness to help, whose integrity is beyond question, a man who understands his most secret and most personal problems, and the Doctor must be such a friend" ([6], p. 73). Physician Thomas P. Almy speaks of the need for the "human touch" and the need for the patient-physician interaction to be "a more personal one" ([1], pp. 403, 407). The same longing for the model of the friend is expressed negatively in criticisms of the bureaucratization of medical care [7], trends toward specialization, and "new intrusions" on the patient-physician relationship from institutionalizing medical care by moving into hospitals, groups, or teams making the patient and the physician "units in a matrix" ([33], p. 44). There can be little doubt that this constellation of personal qualities which I have labelled with the 'friendship' metaphor has been held widely to be an attractive, highly desirable way to have patients and physicians relate to one another.

The Ethics of the Friend/Physician Model

Assuming for the moment that the friend/physician model of the patient-physician relationship is the one that should be normative for medical practice, one might inquire about the relationship of that model to various theories of medical ethics.

The execution of this analysis will require a few words about the notion of 'theories' of medical ethics [46]. It is increasingly becoming apparent that medical ethics is not confined to *ad hoc* moral speculation about difficult cases and difficult topics such as abortion, experimentation, euthanasia, or genetic engineering. There is, and at least implicitly has been for centuries, a deeper, more coherent structure to moral reflection about medical decisions. Each major sociological or sub-cultural group has developed a more or less consistent *Weltanschauung*, a set of metaphysical and metaethical beliefs and values, and a set of basic general principles that inform moral reflections regardless of the topics that are at issue. Roman Catholic moral theology, for example, offers a deep and rich tradition of ethics tracing back at least to Thomas Aquinas. Principles such as the principle of double effect and the principle of totality are derived from a teleological natural law doctrine that informs all work in medical ethics done from within this tradition. (See [12, 22, 24, 25] for a variety of examples of medical ethical scholarship from within this tradition.) Similarly, Jewish Rabbinical scholarship offers a 'theory' of medical ethics closely related to Jewish notions of the law, the sacredness of life, special obligations regarding the burial of the corpse, and the like. (See [5, 19, and 38] for examples of Jewish medical ethics.)

The mainstream of Western professional medical ethics with self-conscious continuity from Hippocrates to the modern Western professional associations of physicians at least up to the middle of the twentieth century also provides a medical ethical theory of substantial consistency. The central ethical principle, as stated in the Hippocratic Oath, is that the physician pledges to work "for the benefit of the sick according to my ability and judgment" ([14], p. 6). There is also a substantial emphasis on the ethics of virtue (although the specific virtues shift from the "purity and holiness" of the Hippocratic author to the "tenderness, steadiness, condescension, authority" of Percival's code published in 1803 [34], p. 71, and finally the modern AMA Principles emphasizing compassion and respect for dignity [2], p. ix).

Although it is not logically required, holders of the Hippocratic normative ethical principle that the physician should do what he thinks will benefit his patient also have a rather consistent view on questions of ethical justification. In response to the question of how one knows that physicians should act on this principle, the normal response of those working within what I am calling the Hippocratic tradition takes one of two forms. Sometimes members of the professional group generate their own moral obligations as part of specifying what it is to be a member of the profession. Alternatively, they say that even though the moral norm is grounded in something more universal (in reason, experience, the laws of nature, or God's will, for example), only members of the professional group can know these principles. Thus, in either case, the professional group is the source of and articulator of the moral norms. For this reason professional groups write and enforce their own codes of ethics.

In contrast a major alternative medical ethical theory seems to be emerging in the late twentieth century influenced by both secular/philosophical and religious (especially Protestant) ethical sources. I and others have been calling this alternative a contract or covenant theory of medical ethics [8, 15, 26, 28, 32, 42, 43 and 46]. While its characterization differs from author to author, its essential characteristics include a broader base for the formulation of medical ethical principles, rules, and duties including significant lay participation in medical ethical reflection; emphasis on deontological duties of fidelity, honesty, and respect for rights; and some notion of contractual or convenantal bond establishing the relationship between the profession and society and, on the individual level, between professional and patient.

What I am now suggesting in this analysis is that the formulation of a model for the patient-physician relationship is influenced by the implicit medical ethical theory under which one is operating. In turn, within any given theory of medical ethics whether the physician is viewed as a friend or as a stranger may influence normative judgments bearing on the patient-physician relationship. For example, if a medical ethical theory included with detail and nuance rules of right conduct for lay-professional interactions (such as perhaps in an extreme form of natural law-based casuistry in some expressions of Catholic moral theology), intimate, long-standing friendship between patient and physician might not be particularly essential *for purposes of determining right conduct* on a whole range of issues such as sterilization, abortion, active killing for mercy, etc. On the other hand, a hypothetical medical ethical theory that operated on the principle that right conduct was determined by having the physician do what was most consistent with the patient's long-term, subjective lifestyle might require a substantial degree of long-term, intimate continuity of the sort I am characterizing by the metaphor, friendship.

The notion of theories of medical ethics gives us an analytical tool for exploring the ethical implications of the friend/physician and the stranger/physician models of the lay-professional relationship. For purposes of simplicity I shall confine this initial exploration to these two major medical ethical theories competing in the ideological marketplace today: the Hippocratic and the contractual theories. We shall see that neither locks one in rigidly to the friend/physician or stranger/physician models. In fact, either can accommodate either of the models, but each requires considerable internal variation and modification to do so.

Hippocratic medical ethics and the friend/physician model. The Hippocratic ethic focuses on the duty of the physician to do what he thinks will benefit the patient "according to his ability and judgment." It is often stated negatively as "First of all, do no harm" and even rendered into Latin as *primum non nocere* (although it is possible that this is a late dignification of a modern folk ethic variant [20, 37 and 46]). The priority of avoiding harm over helping does not exist in any ancient Greek or Latin texts of which I am aware. This Hippocratic tradition (whether stated in terms of positive or negative consequences) shows a particular affinity to the friend/physician model.

The Hippocratic ethic can be characterized as a form of individualistic, paternalistic consequentialism. It is an ethic where consequences are what count. Furthermore the ethic limits the morally relevant consequences to those affecting the patient; it is individualistic. Finally, the standard of reference for judging what will benefit the patient is paternalistic: the benefits are assessed according to the physician's judgment.

In such a medical ethical tradition, which has until recently dominated Western professional physician ethics, the friendship model is almost necessarily of crucial significance. If the physician's task is to use his or her judgment to benefit the patient and one is at all enlightened about the significance of psychological, social, economic, religious, and cultural factors in the patient, then long-standing 'friendship' in which the physician has knowledge of the broadest aspects of the patient's life is virtually essential. It is safe to say that Hippocratic medicine at its best must be holistic, must make the physician, if not a loving father, at least a sibling or 'best friend' in the medical sphere.

There is one way out of the friendship model for the Hippocratic physician. That is by retreating into an objectivist biological reductionism. This would require first of all that the patient's welfare be conceptualized as objective, that is, fixed in reality independent of subjective preferences. More than that is needed, however. If welfare were objective but still highly diffuse (dependent upon psychological, social, economic, and religious realities), the physician would still require diffuse, intimate knowledge of the patient's objective situation. If, however, welfare were reduced to biological welfare (still conceptualized as objective) then quite possibly a physician could do what was in a patient's interest medically without any of the complexities of knowledge about the other spheres of the patient's life, in short, without the friendship. This is precisely what has happened in high technology, bureaucratic, but nevertheless Hippocratic medicine by mid-twentieth century. The physician thought he could be a stranger in some tertiary care center and still 'do what was best for the kidney in room 374.'

High technology physicians are not the only ones who have made the biologically reductionistic move from within an essentially Hippocratic ethic. Ethicist Paul Ramsey, in his recent advocacy of a "medical indications policy" for determining appropriate care for the incompetent terminally ill ([35], p. 178), is committed to determining what is objectively in the patient's interest. He somehow accepts the strange, implausible assumption that what is objectively in the patient's interest can be determined medically by a physician who is a total stranger to the patient, perhaps never knowing that patient while the patient is conscious, or anything about the social system, the family, the psychology, or the belief system of the patient. If the welfare of the patient is objective and can be reduced to the patient's biology, then one can be Hippocratic without the friendship. Otherwise it is hard to see how Hippocratic medicine could be practiced outside the friendship model.

Contract medical ethics and the friend/physician model. The relationship between contract or covenant medical ethics and the friendship model is more complex. Certainly the contract model, properly understood, can easily accommodate the relation of friendship between patient and physician.

One might think that contract implies legalistic or businesslike relation between physician and patient and that the contract model is, therefore, inimical to friendship ([27], p. 33).

No one who takes the contract model seriously, however, has anything like that in mind. If there is to be a contractual *ethic* for medicine, all agree it must avoid those implications, emphasizing instead such contractual notions as moral equality between the partners of the contract, fidelity to promises made, a sense that parties to the contract are autonomous agents capable of pledging and fulfilling pledges. The norms are trust, loyalty, respect, and faithfulness, not legalism and business bargaining. It is the contract of a 'marriage contract', not of the lawyer's office. That is why some now insist upon or at least prefer the term 'covenant', which I take to be one special form of the more general notion of a contract.

The implications of the contract or covenant ethic for the friend/physician model of the patient/physician relationship depends somewhat on the exact working out of the contract theory. I shall illustrate this by using the triple theory of medical ethics, which I have developed elsewhere [44, 45, and 46].

This particular form of contract theory of medical ethics holds there must in fact be three separate contracts. They are taken as ideal or hypothetical contracts. If one wants to know what is really, objectively right morally in medicine, one would have to determine what hypothetical contractors would agree to (or discover or invent, depending on one's metaphysics and theology) under certain ideal conditions (including adequate knowledge, sensitivity, and most significantly taking the moral point of view in which the welfare of all was taken equally into account).

Now, of course, no such ideal contractors exist, but the ideal contract at least gives us an image for conceptualizing what it means to say something is morally right or wrong. Moreover, it gives us a method for real life moral deliberation. Real life contractors come together and try to approximate the conditions of the ideal contractors. The result of their deliberations approximate what the ideal contractors would have agreed upon to the extent that the real life contractors are successful in mimicking the perspective of their fictional counterparts.

In the triple contract theory the first contract would be the most basic social contract establishing the fundamental constitutive moral principles for a society, a moral community. Principles such as fidelity, autonomy, veracity, avoiding killing, justice and beneficence would result.

A second contract would establish a moral relationship between the members of the society and some special sub-groups that are given special moral mandates. The moral framework for professional groups such as physicians would be established at this point. Physicians *and* lay people would have defined for them special sets of rights and obligations. Rights of physicians to control toxic substances, commit 'assault' by cutting the body under special circumstances, and perhaps broad exemptions from general social obligations (such as worrying about broad societal welfare) might be created in this way. Special (if limited) duties of confidentiality, consent, and patient loyalty might be created in this way. The reciprocal sets of rights and duties would apply to professional and lay groups as a whole. The moral characteristics of these groups would be so conceived.

In a liberal society both lay persons and professionals as individuals presumably would retain substantial discretion in many areas of moral action. Physicians and lay people would be free to favor extremely aggressive chemotherapy for terminal cancer or hospice care as their personal systems of beliefs and values dictated. Certain limits would be set. Active killing for mercy might be proscribed (because it violated some fundamental principle of the first contract, for example, a principle against killing or some form of a principle of beneficence such as rule utility [see ([3], p. 188), for the working out of the example]). Substantial choice would be left, however, to individual professionals and lay people. Thus a third contract or covenant between individual physician and patient would establish the unique character of the relationship.

The contract or covenant theory of medical ethics accommodates the friend/physician model as easily as the more traditional Hippocratic ethic does. It does so, however, in a very different way. The Hippocratic ethic requires some degree of what we are calling friendship in order to fulfill the moral mandate to do what the physician thinks will benefit the patient (unless one retreats into the nasty world of an implausible objective biological reductionism). The contract or covenant ethic, as we shall see, does not require friendship, but it can function effectively when the health professional and lay person know each other intimately over a long period understanding each other's beliefs and values.

The contract ethic, when it functions within a friendship model, gives great play and prominence to the third contract. While fundamental principles for social interaction are established in the first contract and the basic nature of the lay-professional relationship has limits set for it in the second, the third contract, the one between individual professional and lay person, will, when they have a long-term, trusting relationship, really be decisive. The physician and patient who know each other's beliefs, values, preferences, and tastes can incorporate them into their mutual understanding of the relationship. If they serve together on their church's 'pro-life' or hospice development committee, they may well already know each other's commitments so that the third contract can be an implicit, presumed agreement affecting many critical medical areas. If the physician sees his patient suffer when agony-prolonging extraordinary care is given to a dying patient, the foundation for spelling out a 'third contract' regarding terminal care is already laid.

If the uniqueness of the beliefs and values of the physician and the patient are to be respected and incorporated into the third contract, some degree of communication about those beliefs and values will have to take place. Over a period of time patient and physician can determine whether there is a sufficient meeting of the minds that a mutually acceptable third contract can be established to provide a basis for the individualized friendship style of health care.

Thus both the Hippocratic and the contract theories of medical ethics are compatible with the friend/physician model of health care. The Hippocratic almost necessarily requires it because the physician, charged with doing what he thinks will benefit the patient, must incorporate knowledge of the patient's life in order to make the judgments called for. The friendship is rather one-sided, to be sure. The patient need know relatively little about the physician's life. He is a rather passive partner in the friendship. An ongoing, trusting relationship is, however, essential. The covenant or contract ethic can accommodate friendship in a truer, more reciprocal sense. Each needs to know the beliefs and values of the other to determine the content of the third contract, that is, to decide whether the commitments can be fine-tuned successfully to establish a unique relationship.

THE STRANGER/PHYSICIAN AS THE REALITY

The reality of the delivery of health care in most industrialized societies places severe constraints on the possibility of the development of the friend/physician model. A substantial portion of health care today is delivered in institutional settings where ongoing relationships or friendships are extremely difficult if not impossible. The stranger/physician relationship is the norm in urban clinic and hospital outpatient services, student health services, military and veteran's hospitals, as well as tertiary care settings and specialist referrals. It is extremely difficult to determine exactly what percentage of patient/physician relationships are anonymous, that is, between strangers. Some sense of the growing importance of anonymous contacts can be gleaned from the national and local data.

The 1978 National Health Interview Survey of the civilian, noninstitutionalized population revealed that, on an annual basis, for every 1000 people 642.7 physician visits took place in hospital outpatient departments and emergency rooms ([40], p. 168). By contrast 3158.3 took place in doctors' offices or clinics or group practices. It is safe to assume a substantial portion of the outpatient and virtually all the emergency room contacts were anonymous. Some smaller percentage of clinic and group practice contacts are also anonymous, and even some contacts in solo practice private offices are. Furthermore, these data exclude noncivilian care and care of 'institutionalized patients' (apparently referring to prisons, mental institutions, and other facilities where continuity of care is often minimal). Even among visits to office-based physicians, only 56.5 percent are to primary care physicians (GPs, internists, and pediatricians) ([41], p. 4). Even in a major university family practice center where continuity of care was a central commitment and patients were assigned to a specific physician, the average level of continuity of care for primary care was only 81 percent ([36], p. 529). Given the mobility of the American population and the high incidence of 'doctor-shopping' by dissatisfied patients (between 37 and 48 percent within a year depending on income level, in one study) ([21], p. 328), combined with the increasing incidence of institutionally-based care, it is crucial to challenge the presumption that the ethics of the professional health care relation should be premised on the friend/physician model. It is critical to explore how the major theories of medical ethics can cope with the stranger/physician model.

Hippocratic Medical Ethics and the Stranger/Physician Model

I have already argued that the Hippocratic medical ethic is in serious trouble when the relationship between the patient and the physician is an anonymous one. In fact it may not be an exaggeration to say that it is *impossible* to fulfill the Hippocratic ethical mandate when the patient and physician are strangers.

Aside from the move to biological reductionism, which is so implausible that it does not warrant further attention, there is simply no way for the stranger/physician to know what is really in the patient's welfare. There is simply no information upon which the physician can base his or her judgment.

In circumstances when a misguided physician tries to continue in the Hippocratic mode even though the patient is a stranger, tragedy is likely to result. Dr. Robert Morse, the physician randomly assigned to Karen Quinlan, tried to guess at what was in her interest in spite of the fact that he literally had never met her while she was conscious. His uninformed guess about her welfare turned out to be sadly out of line with more informed judgments of her family, her friends, her priest – the unanimous judgment of all who knew her. The greatest danger of the Hippocratic ethic is that the friend/physician model may be assumed when the stranger/physician is the reality. If the physician must provide care for a stranger, it simply must be done on some basis other then Hippocratic paternalism that requires, at minimum, knowledge of the patient.

Contract Medical Ethics and the Stranger/Physician Model

It seems as though the stranger/physician relationship is also less than optimal under a contract or covenant ethic. In the language of the contract model developed here, when patient and physician are strangers there can be no third covenant – no sharing between lay person and professional of an understanding of unique beliefs and values.

In the case of the contract ethic, however, medical interaction when the physician and patient are strangers is still possible. The moral structure is not limited to individualized judgments about patient welfare by individual physicians who have to guess about what counts as patient welfare given the objective and subjective social, psychological, economic, religious, and other valuative dimensions of the patient's life. The care is delivered in the framework of the two prior contracts. When the relationship is anonymous, the relative significance of the contracts simply shifts. The third contract decreases in importance. In the extreme case, when the patient is unconscious, for example, it disappears altogether. The second contract (between the professional and the society as a whole) remains to define the moral structure of the care to be delivered.

In a world of anonymous health care between people who are essentially strangers the second contract will have to be developed to signal to all involved, patterns of interaction that are to be presumed until such time that the physician and patient (or agent for an incompetent patient) reformulate them by forming a limited third contract.

For example, consider the doctrine of the so-called therapeutic privilege ([47], pp. 8, 17–20; [29], pp. 460–470). Under this doctrine physicians claim the 'privilege' of withholding relevant information from patients such as the diagnosis of a terminal illness, the risk of a therapeutic procedure, or even the entry of the patient into a research protocol. The information is withheld on the grounds that disclosure would be so upsetting for the patient that on balance it is in the patient's interest not to have it.

The therapeutic privilege is made for the Hippocratic ethic. Disclosure and nondisclosure are dictated by the moral mandate to do what the physician thinks will benefit the patient based on the physician's judgment. In the ideal world of the friend/physician model, the physician, through long, intimate contact with the patient, would know precisely what information the patient would want to know and what he would want to have withheld on the grounds that it would be too upsetting. If the goal of the therapeutic encounter is paternalistic patient benefit and the model is friendship, the therapeutic privilege might be defensible.

In the contract ethic a radically different approach is taken toward the therapeutic privilege. If the relation is one of friendship and the third contract dominates, patient and physician can agree together whether or not the physician (or anyone else) should have the authority to withhold information on paternalistic grounds. If patient and physician over the years have a clear understanding about what kinds of information would be upsetting and the patient has agreed that his physician should treat him paternalistically regarding withholding information, then the therapeutic privilege is probably tolerable. It is not a very dignified way to treat one's friends, but in certain relationships paternalistic withholding might be understandable. Likewise, patients might paternalistically withhold from their physicians information that they think might upset them – that for rational reasons a second opinion has been sought, that a medication has been discontinued, or that some element of the physician's practice is slightly offensive. The danger of such paternalistic agreements, of course, is that one of the friends might guess incorrectly about what is really in the other's interest. It is hard to argue, however, that limited agreements authorizing mutual paternalism should be made illegal among consenting adults.

Likewise physicians and patients who know each other over a period of time can negotiate a third contract excluding such paternalism as the therapeutic privilege. Some patients, probably most in today's world, would not want information withheld on traditional therapeutic privilege grounds. Many physicians are also finding that the stance of the therapeutic privilege is uncomfortable or even offensive. If patients and physicians who share opposition to the therapeutic privilege or want to limit it in certain ways wish to agree to such a lay/professional relationship, under the contract theory this is morally the appropriate way to handle the problem.

In the physician/stranger model there is no basis for the therapeutic privilege in the Hippocratic ethic. There is no reason for the physician to know what information would upset the patient. The professional is reduced to the dangerous 'golden rule' type of reasoning where the physician's own feelings, psychological set, and predicted reaction to bad news is substituted for that of the patient.

When patient and physician are strangers under a contract ethic there is still no opportunity to fine tune the doctrine of therapeutic privilege using what we are calling third contract. Some pattern must be established at the level of the relationship between the society and the profession, that is, the level of the second contract. The two groups coming together could agree that in stranger relationships professionals will use their instincts to withhold information on therapeutic privilege grounds. It seems far more plausible, however, that the society and the profession would establish a fall-back position that stranger-patients should be told what reasonable people would want to know about their situation *even if the physician believes this might upset them*. This is precisely the position the society is adopting in the 'reasonable person' court cases [4, 9, 10, 11, 13, 18, and 49].

Recently the question has arisen of whether the reasonable person standard requires that patients be told what the typical 'objective' reasonable person would want to know or whether the information should be adjusted to take into account the special subjective desires, life-plans, and values of the specific patient ([30], pp. 287–288; [39]). The answer that seems to be emerging is that, with certain qualification [48], patients should be told what the typical 'objective' reasonable person would want to know in order for a consent to be adequately informed unless the one obtaining the consent has reason to believe that the specific patient would desire to know more or less in certain areas.

This recent development can be understood easily in the framework of the models of physician/friend and physician/stranger. In the pure case of the physician/stranger (where the physician knows nothing about the uniqueness of the patient) the 'objective' reasonable person standard applies. The second contract in effect requires that physicians tell patients what the objective reasonable person would want to know in cases where no unique information is known about the patient. As the physician gains knowledge of the specific patient and his or her unique interests, the friendship model begins to become more appropriate. When the physician has reason to believe that the patient would reasonably require special kinds or amounts of information, the third contract requires that the more subjective reasonable person standard be substituted. The patient bears substantial responsibility for modifying the understanding of what a patient would want to know (although the physician may have the obligation to make reasonable deductions about the patient's uniqueness from the information available and to point out to the patient that his or her unique interests may be important in deciding what to disclose). The medical profession is increasingly accepting the wisdom of this position as well. In effect the second contract, the one between the profession and the society, establishes a fall-back position on the doctrine of therapeutic privilege: it is unacceptable unless physician and patient agree between themselves on an individual basis for some authorization of it.

The result is that the second contract establishes a societal, ethical pattern for all relationships between professionals and lay people that is publicly known and applicable to all unless modification is made between individuals who hold some other set of beliefs and values.

A wide range of problems in medical ethics can be handled in this way. Suppose a woman has been raped and seeks medical attention. Should the physician take the initiative to recommend initiation of diethylstilbestrol, which would function as a morning-after abortifacient? Some physicians may find even very early abortion following rape morally unacceptable. Furthermore some physicians may make the judgment that for particular patients morning-after abortifacients may have serious psychological consequences. Given the wide range of physician beliefs and interpretations of potential consequences some small percentage are likely to determine that it is morally unacceptable or contrary to the welfare of the patient even to initiate such conversation. Yet surely many such patients, especially those who were not aware of such a pharmacological option, would want the conversation initiated by the physician.

Under the Hippocratic ethic of paternalistic patient benefit, the physician must use his or her judgment in deciding whether to initiate. If the patient and physician have a long term friendship, perhaps knowing each other through religious group as well as medical relationship, the physician may well guess correctly. In fact some woman might be offended to have her physician suggest an abortifacient. If they are strangers the physician can only guess.

Under the contract ethic a community would mandate under the second contract a pattern of standard practice that both physician and patient could come to expect.

The most obvious second contract standard policy might be that physicians doing rape-crisis medical counseling should initiate such conversation until such time as the patient signals the discussion should cease. Patients who were strangers would indicate that this was a course they were not willing to pursue, if that were the case. The physician who, for moral or other reasons, could not tolerate having to initiate such conversations would withdraw from rape crisis medical services or limit such practice to patients with whom at least enough of a relationship had been established that a more individual (third contract) agreement was established that morning-after abortifacients following rape were excluded from the agenda. Once again the key is that with a contract ethic, some standard pattern of ethically acceptable behavior for lay people and professionals is established that sets a predictable pattern for stranger relationships. If the patient/physician relationship had evolved to the point where some personal relation were established, special modifications could be negotiated to permit deviations from the standard provided they did not violate more basic moral or legal obligations of the community.

The contract or covenant ethic is equipped to handle friendship relations when they exist. It is also, however, appropriate for structuring basic moral patterns for relationships between strangers that are increasingly the statistical norm.

THE STRANGER/PHYSICIAN AS IDEAL

The increasing statistical frequency of the stranger relationship between physician and patient, of course, in no way implies that the relationship is morally normative. In fact it is widely believed that the stranger/physician is at best an unfortunate necessity in a world of bureaucratic health care and a mobile society. Thus far the analysis of the implications of alternative medical ethical theories for the friendship and stranger relations has been descriptive. The goal has been to see how far each theory could take us in understanding the ethical dynamic of health care among those who are friends or strangers. It is interesting, however, to pursue a more normative assessment of the physician as stranger. This final section will attempt to initiate such a pursuit.

I take it as an obvious starting point that many people see the important advantages of the friendship model. All of us would like to be able to fine-tune the relationship with health professionals so that it takes into account our unique beliefs, values, ethical commitments, and lifestyle. Some degree of intimacy is required for this to happen, even if it is only making sure to your oncologist that you are the type of person who wants to fight the cancer to the last gasp rather than be transferred off to a hospice. I shall take the advantages of the friendship model as more or less obvious, the consensus wisdom of the community.

The interesting question is whether there is any reason why some people might actually prefer the physician as stranger, at least for certain kinds of medical care encounters. To initiate the catalogue of possible reasons for preferring the stranger/physician it is apparent that many patients seem to prefer the stranger in certain highly specialized, sensitive kinds of care. Patients seeking abortions, venereal disease treatment, and, possibly, mental health therapy, often seem to seek out strangers wanting the anonymity that comes with it rather than the burden of having to discuss such matters with physicians who are long-term acquaintances. The anonymous patient-physician relationship offers a protection of privacy not available with more personal professional relations.

More generally, some people seem to have personality types that favor maintenance of more compartmentalized lives. They simply would rather lead their lives in such a way that those they know in one sphere are not deeply involved in other spheres, that those they see in church are not the same ones they deal with in business, in their children's school activities, in their entertainment, or their health care. One type of contemporary personality seems not to require integration of the spheres of life. It is a pattern the psychiatrist Robert Lifton has labelled the 'protean man' [23]. Some people would just prefer that they not have to spend time discussing their children with the grocery store clerk. Possibly these same personality types would rather not have to develop the more sociologically diffuse type of relationships with physicians. The reasons for this are numerous. Some, perhaps, prefer not to have their 'friends' see them in weakness, which often (but not always) is the condition of patients seeking medical professional help. Others may simply prefer the sense of mystery that comes from being more private people or simply not care to spend the time and energy necessary to build this particular set of friendships.

Another potential advantage of the stranger/physician model is that it avoids the danger of having the friend/physician attempt to estimate patient welfare or preferences based on what may be inadequate knowledge of the patient. Both the Hippocratic and the covenant medical ethics leave room for the physician to make judgments based on their perception of the patient's subjective position. The Hippocratic does this more openly in not only authorizing but actually encouraging paternalism. The contract ethic leaves open the possibility that patients and physicians will gradually come to believe they have an understanding of a special set of normative relationships – of a unique third contract modifying the more general second contract. If one is concerned that there is a danger of misunderstanding in such presumed third contracts, then one might actually prefer the stranger relationship where no presumption could plausibly be warranted.

Upon reflection it appears that one's preference for the stranger or friendship model of health care will depend on many variables including personality type of patient and physician, the nature of the medical problem under consideration, the type of medical specialty involved, and so forth. It may also depend on how close one feels he or she is to what might be called the 'core values' of the culture. If one is close to those core values, one is likely to look approvingly on the general conditions of the second contract so that one's personal values will be reflected in the standard ethical presumptions the second contract dictates for the stranger/physician. On the other hand, if one's personal values deviate greatly from the 'core values' one might make a special effort to seek out in advance a physician who can get to know the patient's uniqueness and accept or at least tolerate it. On the other hand the person whose personal values differ widely from the core values may fear that a friend/physician may continue to act on the old Hippocratic ethic thus acting paternalistically doing what the physician thought would really serve the patient rather than what the patient preferred at the time. He or she may purposely seek out the stranger knowing that the stranger/physician ought to be more reserved in his paternalism and more confined by the second contract. The person with unique values would then carry the burden of informing the stranger/physician of his uniqueness, but would at least have the comfort of knowing that such a stranger/physician ought to behave in an ethically predictable manner and be open to patient-initiated discussions of variations. The stranger/physician may be less offended or threatened than the friend/physician when the patient presses his or her uniqueness.

One final argument in defense of the stranger relationship is quite subtle. It might be argued that patient dignity is better preserved in the stranger relationship than in the friendship model. That certainly would not always be the case, but may accurately describe a tendency. For example, it might be felt that equal respect or dignity is best preserved between adults (age and other variables being equal) when each uses the same degree of familiarity in referring to the other – that both are on a first-name basis or refer to each other using surnames. While in the friendship relation occasionally patients may address their physicians on a first-name basis, it is probably typical that the physician will be referred to by title and surname while the patient is often addressed on a first-name basis. In the stranger relationship, on the other hand, both parties may be inclined to use surnames. Increasing familiarity in a sociological context of substantial status differential may increase the gap in the degree of respect shown between the parties.

Certainly the stranger/physician is not for everyone or for all conditions. Certain types of medical practice may require the friendship or stranger relation more than others. For example, there is one type of practice where the stranger relationship is normative – actually preferred as the ideal. In psychiatry establishing a therapeutic relationship with an old friend would be frowned upon. It is often considered especially important that the personal beliefs and values of the psychiatrist not be known to the patient. This is particularly strange since if there is one sphere of medical practice where the friendship model is especially important to patients it would have to be psychiatry. Major therapeutic strategy choices and even the choice of underlying theory are dependent upon the normative system of beliefs and values of the psychiatrist. If the patient comes to the psychiatrist, for example, with a problem of a behavior that produces guilt, there is no way for the psychiatrist to know whether to treat the behavior or the guilt without deciding whether the behavior is appropriate and there is no way of deciding that without a set of ethical and other normative commitments. One of the great differences between a psychiatrist and a minister/pastoral counselor is that the minister/pastoral counselor has stood before his or her potential client week after week testifying to the personally held theological, ethical, and philosophical beliefs that are certain to influence the counseling.

In more ordinary kinds of medical care, however, some people may actually prefer to take their chances with the anonymous relationship provided it is properly structured ethically by societal level moral structures that we have been calling the second contract.

The implications for medical ethics of the identification of the unique ethical character of the anonymous patient/physician relationship are enormous. If the Hippocratic ethic is to be restored, the friendship model seems crucial. To the extent that more anonymous relationships are either inevitable or desirable, defenders of the Hippocratic ethic must look to the future with despair.

If the future includes within it medical practice by algorithm and eventually by computerized diagnosis and even mechanized store-front diagnostic centers (the grocery store coin-operated blood pressure machines being the precursors), then some other medical ethical theory will be a necessity. The contract or covenant theory is an alternative. It can accommodate the friendship model quite nicely, but also has the potential for providing a medical ethical foundation for health care relations among strangers. If it turns out that the stranger/physician is actually preferable, at least for certain people and certain kinds of care, then the contract ethic approach or some surrogate of it, will be essential. It provides moral frameworks for both friendship and stranger relationships although those frameworks are rather different.

Kennedy Institute of Ethics,
Georgetown University, Washington, D. C.

BIBLIOGRAPHY

1. Almy, T. P.: 1980, 'The Healing Bond', *The American Journal of Gastroenterology* 73, 403–407.
2. American Medical Association: 1981, 'Principles of Medical Ethics', in *Current Opinions*, Judicial Council, American Medical Association, Chicago.
3. Beauchamp, T. L.: 1979, 'A Reply to Rachels on Active and Passive Euthanasia', in W. L. Robison and M. S. Pritchard (eds.), *Medical Responsibility: Paternalism, Informed Consent, and Euthanasia*, The Humana Press, Clifton, N. J., pp. 182–94.
4. *Berkey v. Anderson*, 1 Cal. App. 3d 790, 805, 82 Cal. Rptr. 67, 78 (1969).
5. Bleich, J. D.: 1979, 'The Obligation to Heal in the Judaic Tradition: A Comparative Analysis', in F. Rosner and J. D. Bleich (eds.), *Jewish Bioethics*, Sanhedrin Press, New York, pp. 1–44.
6. Bleuler, M.: 1973, 'Let Us Stay Near Our Patients', *Diseases of the Nervous System* 34, 73–79.
7. Bloom, S. W., and Summey, P.:1977, 'Physician-Patient Expectations in Primary Care', *Bulletin of the New York Academy of Medicine* 53, 75–82.
8. Brody, H.: 1976, 'The Physician-Patient Contract: Legal and Ethical Aspects', *The Journal of Legal Medicine* 4 (July–August), 25–29.
9. *Canterbury v. Spence*, 464 F2d 772 (D.C. Cir.), cert. denied, 409 U.S. 1064 (1972).
10. *Cobbs v. Grant*, 502 P. 2d 1 (Cal. 1972).
11. *Cooper v. Roberts*, 286 A. 2b 647 (Pa. 1971).
12. Curran, C. E.: 1979, 'Roman Catholicism', in W. T. Reich (ed.), *Encyclopedia of Bioethics* 4, The Free Press, New York, pp. 1526–1530.
13. *Dow v. Kaiser Foundation*, 90 Cal. Rptr. 747 (Cal. 1970).
14. Edelstein, L.: 1967, *American Medicine: Selected Papers of Ludwig Edelstein*, Johns Hopkins University Press, Baltimore.
15. Epstein, R. A.: 1976, 'Medical Malpractice: The Case for Contract', *American Bar*

Foundation Research Journal, 87–149.
16. Flathman, R.: 1982, 'Power, Authority, and Rights in the Practice of Medicine', in G. Agich (ed.), *Responsibility in Health Care*, D. Reidel Publishing Company, Dordrecht, pp. 105–125.
17. Fox, R. C., and Lief, H.: 1963, 'Training for "Detached Concern" in Medical Students', in H. Lief *et al.*, *The Psychological Basis of Medical Practice*, Harper and Row, New York, pp. 12–35.
18. *Hunter v. Brown*, 484 P. 2d 1162 (Wash. 1971).
19. Jacobovits, I.: 1959, *Jewish Medical Ethics*, Block, New York.
20. Jonsen, A. R.: 1977, 'Do No Harm: Axiom of Medical Ethics', in S. F. Spicker and H. T. Engelhardt, Jr. (eds.), *Philosophical Medical Ethics: Its Nature and Significance*, D. Reidel Publishing Company, Dordrecht, pp. 27–41.
21. Kasteller, J. *et al.*: 1976, 'Issues Underlying Prevalence of Doctor-Shopping Behavior', *Journal of Health and Social Bahavior* 17, 328–339.
22. Kelly, G.: 1958, *Medico-Moral Problems*, The Catholic Hospital Association of the U. S. and Canada, St. Louis, pp. 1–16.
23. Lifton, R.: 1971, 'Protean Man', *Archives of General Psychiatry* **24**, 298–304.
24. McCormick, R. A.: 1981, *How Brave a New World: Dilemmas in Bioethics*, Doubleday and Company, Garden City, N. Y.
25. McFadden, C. J.: 1967, *Medical Ethics*, F. A. Davis Co., Philadelphia.
26. Magraw, R. M.: 1973, 'Social and Medical Contracts, Explicit and Implicit', in R. L. Bulger (ed.), *Hippocrates Revisited: A Search for Meaning*, Medcom Press, N. Y., pp. 148–157.
27. Masters, R.: 1975, 'Is Contract an Adequate Basis for Medical Care?' *Hastings Center Report* **5** (December), 24–28.
28. May, W. F.: 1975, 'Code, Covenant, Contract, or Philanthropy', *Hastings Center Report* **5** (December), 29–38.
29. Meisel, A.: 1979, 'The "Exceptions" to the Informed Consent Doctrine: Striking a Balance Between Competing Values in Medical Decisionmaking', *Wisconsin Law Review* **2**, 413–465.
30. Meisel, A. *et al.*: 1977, 'Toward a Model of the Legal Doctrine of Informed Consent', *American Journal of Psychiatry* **134**, 285–289.
31. Parsons, T.: 1964, *The Social System*, The Free Press, New York.
32. Pellegrino, E.: 1973, 'Toward an Expanded Medical Ethics: The Hippocratic Ethic Revisited', in R. J. Bulger (ed.), *Hippocrates Revisited: A Search for Meaning*, Medcom Press, New York, pp. 133–147.
33. Pellegrino, E.: 1975, 'Protection of Patients' Rights and the Doctor-Patient Relationship', *Preventive Medicine* **4**, 398–403.
34. Percival, T.: 1927, *Percival's Medical Ethics*, C. D. Leake (ed.), Williams and Wilkins, Baltimore.
35. Ramsey, P.: 1978, *Ethics at the Edges of Life*, Yale University Press, New Haven.
36. Rogers, J. and Curtis, P.: 1980, 'The Achievement of Continuity of Care in a Primary Care Training Program', *American Journal of Public Health* **70**, 528–530.
37. Sandulescu, C.: 1965, '*Primum non nocere*: Philological Commentaries on a Medical Aphorism', *Acta-Antiqua Hungarica* **13**, 359–388.
38. Siegel, S.: 1978, 'Medical Ethics, History of: Contemporary Israel', in W. T. Reich (ed.), *Encyclopedia of Bioethics* **2**, The Free Press, New York, 895–896.

39. Strong, C.: forthcoming, 'Full Competence, Self-Determination and Informed Consent', *Values and Ethics in Health Care*.
40. U. S. Department of Health and Human Services: 1980, 'Use of Ambulatory Care by the Poor and Nonpoor', in *Health: United States, 1980*, National Center for Health Statistics, Publication No. 81–232, December.
41. U. S. Department of Health and Human Services: 1981, *1979 Summary, National Ambulatory Medical Care Survey, U. S., Jan.–Dec. 1979*, by T. McLemore, Vital and Health Statistics. Advanced Data Report No. 66, National Center for Health Statistics, Publication No. (PHS) 81–2150, March.
42. Veatch, R.: 1971, 'Value-Freedom in Science and Technology', Doctoral Dissertation, Harvard University, Cambridge, Mass.
43. Veatch, R. M.: 1972, 'Models for Ethical Medicine in a Revolutionary Age', *Hastings Center Report* 2 (June), 5–7.
44. Veatch, R. M.: 1979, 'Professional Medical Ethics: The Grounding of Its Principles', *The Journal of Medicine and Philosophy* 4 (March), 1–19.
45. Veatch, R. M.: 1981, 'Federal Regulation of Medicine and Biomedical Research: Power, Authority, and Legitimacy', in S. F. Spicker *et al.* (eds.), *The Law-Medicine Relation: A Philosophical Exploration*, D. Reidel Publishing Co., Boston, pp. 75–91.
46. Veatch, R. M.: 1981, *A Theory of Medical Ethics*, Basic Books, Inc., New York.
47. Veatch, R. M.: 1981, 'When Should the Patient Know?' *Barrister* 8, 6–8, 17–20.
48. Veatch, R. M.: forthcoming, 'Consent, the Individual, and the Reasonable Person: Commentary on Carson Strong', *Values and Ethics in Health Care*.
49. *Wilkinson v. Vesey*, 295 A. 2d 676 (R.I. 1972).

JOHN LADD

THE INTERNAL MORALITY OF MEDICINE: AN ESSENTIAL DIMENSION OF THE PATIENT-PHYSICIAN RELATIONSHIP

Discussions of the ethical aspects of the physician-patient relationship generally concentrate on interpersonal questions about such things as decision-making, mutual rights and responsibilities, autonomy and paternalism, treating patients as persons rather than as cases, truth-telling, and informed consent; or else they are concerned with more specialized topics such as euthanasia, abortion, genetic counselling, refusing treatment, orders not to resuscitate, and triage. All of these questions are obviously important. In this essay, however, I propose to approach the physician-patient relationship somewhat differently, namely, from the point of view of medical professionalism. I shall examine the ethical requirements and constraints governing this relationship that are binding on physicians by virtue of their status as members of the medical profession. The body of norms that are involved here will be referred to as *the internal morality of medicine.*[1]

When we talk with physicians about problems of medical ethics, more often than not, they appeal to norms, rules and principles that come from the internal morality of medicine; for obvious reasons, they find it easiest to resort to ethical conceptions drawn from medical professionalism. Most laymen, including patients, are either unaware of or else do not place much credence in norms of this kind. Instead, they are likely to invoke general ethical conceptions accepted by society at large or by its educated and intelligent members. The result is that there is often an impasse between them and physicians concerning the rights and wrongs of the physician-patient relationship. I attribute the breakdown to the fact that insufficient attention has been paid by all parties to the internal morality of medicine, its ethical status, its rationale, and perhaps also, its inadequacies.

An example of the way in which conceptions of what is morally required of physicians depend on differences of perspective of the kind just mentioned is the current debate over the difference between 'killing' and 'letting die', that is, the distinction between what is sometimes called 'active' and 'passive' euthanasia. Philosophers have generally attacked the distinction as a distinction without a difference, arguing, for example, that letting a person die is a way of killing him and that, when one comes down to specific 'doings', any attempt to draw the line between the two is quite arbitrary.[2] Nevertheless,

Earl E. Shelp (ed.), The Clinical Encounter, 209–231.

when viewed from the point of view of the internal morality of medicine and the accepted norms of medical practice, the distinction begins to make sense. For it can be given a meaning in relation to certain widely accepted norms of medical practice according to which a physician is permitted (or perhaps even required) to refrain from further treatment of a patient if the treatment is futile and the patient's case is hopeless, but according to which, on the other hand, he is required to refrain from doing anything to or for a patient that does not have a clear medical purpose. 'To let a patient die' simply means to comply with the first of these norms, whereas to 'kill him', e.g., by injecting a death-inducing drug, would be a violation of the second. The important point to note here is that the distinction is dependent on the norms just mentioned and that logically speaking the distinction itself is derived from the norms rather than the other way around. In order to understand and to apply the distinction, one must therefore have some prior conception of what is required or forbidden by these norms. Different interpretations of what is to count as killing or as letting die simply reflect different conceptions of what it is proper or improper for a physician to do. Like other terms in our ethical vocabulary, such as 'murder', values (norms) determine the facts as much as the other way around.[3]

When I call the body of norms that come from medical professionalism an internal *morality*, I use the term 'morality' advisedly. For the norms in question, even though they are frequently held to be 'medical' rather than 'ethical', actually have moral content: they make moral claims and have moral force. They have moral implications and moral consequences, as we shall see.[4]

In order to make clear the kind of norm that I have in mind, it will be helpful to begin with a little list of some of the norms that might be included under the internal morality of medicine. At this point, I shall simply describe the norms without examining their rationales or their ultimate ethical acceptability.

At the top of the list, there is an array of professional norms condemning unnecessary interventions: unnecessary tests, X-rays, biopsies, and surgery. Then, there are norms condemning unorthodox treatment, e.g., the use of Laetrile, and condemning the use of certain medical procedures for non-medical purposes; e.g., the use of amniocentesis solely in order to determine the sex of a baby. There are also, of course, norms relating to the prescription of drugs, e.g., the overuse of penicillin or the excessive prescription of tranquillizers. Norms like these define the difference between 'professional' and 'unprofessional' medical practice, although, as I have just said, they may also be regarded as 'moral' or 'quasi-moral' norms. For, whatever they are in particular, these norms make a physician 'less free to choose than the patient' and they represent a physician's 'responsibility to the medical enterprise'.[5]

Needless to say, norms such as these are not accepted by all physicians, although the ones just mentioned may seem obvious and uncontroversial. Some of the norms that might be included in an internal morality of medicine are very controversial. Leon Kass, for example, takes a rather extreme view of what is and what is not proper for a physician to do, basing his contentions on what he claims to be the 'true' goal of medicine, the pursuit of health. He asserts that physicians serve 'false' goals if they perform artificial insemination, vasectomies, and abortions for non-medical reasons ([10], p. 5). William Curran also appeals to the goals of medicine to support his contention that physicians should not participate in capital punishment, e.g., in preparing and administering lethal drugs to condemned murderers, or in other activities such as those involving torture or the interrogation of prisoners [5]. Some physicians argue that physicians ought not to cooperate in planning for nuclear warfare, because the effects of a nuclear war would be so disastrous from a medical point of view.

It should be observed that neither Kass nor Curran base their conclusions on the premise that the acts in question are wrong in themselves, e.g., for anyone; they simply argue that it is wrong for physicians to do these things. There is, however, one important logical difference between the arguments given by Kass and by Curran. Kass argues that physicians ought not to do certain things because they *do not further* the aims of medicine; Curran, on the other hand, argues that they ought not to do certain things because they are *inconsistent* with the aims of medicine. Arguments of the latter kind are more cogent logically than those of the former kind. The other norms mentioned may, of course, also be related to some conception or other of the aims of medicine. The question of how this is done will be discussed later.

A few general points about the norms of the internal morality of medicine are in order. First, the norms are thought to be binding on all physicians alike, although perhaps not on others who are not physicians. In this sense, the norms in question might be said to be *sui generis*; for they obviously differ in important respects from the norms that patients and laypersons in general might accept as valid or, for that matter, that they might even think of. Second, physicians who disagree violently about other fundamental ethical or social questions are apt to agree about professional norms. There usually is, and almost always is assumed to be, a *consensus* about them. Indeed, physicians are generally so completely convinced by the norms of the internal morality of medicine and so deeply committed to them that they are often not even aware that they are ethical norms, much less that they might be challenged and criticized ethically. Instead, they speak of them as 'medical norms', sometimes even assuming them to be 'scientific'. By thus isolating and insulating them from ethics of other kinds, physicians are able to claim that only physicians can be experts in medical ethics, for only they can know what the norms of medical practice, the internal morality of medicine, prescribe.

THE INTERNAL VS. THE EXTERNAL MORALITY OF MEDICINE

The internal morality of medicine must be distinguished from what will be called the *external morality of medicine.*[6] Both moralities relate to ethical issues encountered in medical practice and in clinical medicine, but they differ in how they approach them and in their rationales. Under the external morality of medicine I include moral considerations that come from outside medicine and that are based on external non-medical facts such as particular social conditions, the personal idiosyncracies and demands of individual patients and their families, the social and institutional milieu of treatment, and so on. Internal morality, as I have already indicated, comprises moral norms relating to the clinical situations that depend on 'medical' considerations, such as diagnosis, prognosis, treatment plans, concepts of disease, and so on.

The difference between the internal and the external moralities of medicine can easily be seen if we consider the case of the Jehovah's Witness who refuses to have a blood transfusion on religious grounds. From the medical point of view, that is, from the point of view of the internal morality of medicine, a transfusion is (morally) required, for it is necessary in order to save the individual's life. From the external point of view, however, it would be morally objectionable, since to give the individual a blood transfusion would require him 'to eat blood' and thus, in his eyes, to violate God's law. We do not have to accept the religious reasoning in order to accept the principle that it would be wrong to force a person to accept treatment that goes against his conscience. The right to refuse treatment and other issues involving patient's choices, whether justified or not, fall under the external rather than the internal morality of medicine.

Telling the truth to patients is a general requirement of the external morality of medicine. Sometimes it is inconsistent with the internal morality of medicine, as, for example, when physicians invoke the so-called 'therapeutic privilege'. More often than not, it is consistent with both moralities and sometimes it is required in an individual instance by both of them. It should be observed, however, that they use different kinds of reasons. The external morality appeals to the precept of honesty, whereas the internal morality focusses on what is best for a patient's recovery.

The way decisions and other sorts of actions are categorized morally often differs radically when they are considered from the point of view of the two moralities. Indeed, the categorization is often cross-wise; that is, the internal morality may prescribe the same treatment for two different patients with the same disease and the external morality may prescribe treating them quite differently; for one patient may desire to be treated and the other may refuse. Analogously, the same external considerations may prescribe the same kind of action (or non-action) with respect to patients with quite different medical requirements; for example, a physician may be required by the external morality to refrain from treating two patients, one of whom has an easily treatable disease but refuses treatment and the other of whom refuses but who has an incurable, terminal disease that ought not to be treated anyway, from the medical point of view and from the point of view of the internal morality of medicine. The fact that in cases like these we have to cope with differential and cross-wise classifications simply bears witness to the complicated character of morality in general and of the decision-making it pertains to.

It is obvious that it will not always be possible to draw a sharp line between what is required by the internal morality and what is required by the external morality of medicine; for, in a deeper sense, all the relevant factors in a medical situation are interconnected and interrelated. The fact, for example, that a patient is an alcoholic may function as both an external and an internal consideration in determining what kind of treatment he should receive. I have already given examples where the same kind of action, e.g., truth-telling, may in particular circumstances be required by both. Even more puzzling are those cases where it is unclear what is required by one of the moralities, such as the use of narcotics as pain-killers.

A few general remarks about these two moralities should put my contentions about them in proper perspective. To begin with, like a great deal of morality in general, conclusions about what is or what is not part of one of these moralities are controversial. That is not to say, however, that they cannot be rationally appraised and criticized or that they are incapable of rational resolution. Here again, I suggest that we try to deal with specifics rather than generalities. If we admit that persons may be mistaken in their ethical opinions, then we must also admit that it is possible to criticize them and to attempt to correct and improve them. This is the supposition that I shall take for granted in my later discussion of the particulars of the internal morality of medicine.

Second, as with all moral norms, that is, rules and principles, the norms that concern us here include a *ceteris paribus* clause: they are binding only as long as other things are equal. Other things being equal, one should strive to keep a person alive; or other things being equal, one should respect a person's refusal of treatment, and so on. All the requirements contained in the norms in question are, in philosophical jargon, *prima facie* duties.[7] Of course, things are often not equal and we are confronted with different norms (*prima facie* duties) prescribing opposite and incompatible courses of action and are forced to choose between them. In such cases, which are not infrequent in the medical world, the physician and others who are involved face a moral dilemma – an either/or situation in which one has to decide which of the alternatives ought to be given preference. There is no simple rule that will help us to resolve moral dilemmas in general, although often the choice is obvious. Sometimes medical considerations override external moral ones and sometimes, of course, it is the other way around.

When faced with dilemmas like these, we must appeal to the maxim: *ought implies can*; for no one is required to do what is impossible and it is impossible in a dilemma to satisfy both horns. It is impossible, for example, both to save the Jehovah Witness's life and to honor his right to refuse a transfusion. One has to choose and choosing involves deciding which, under the circumstances, is the better or more reasonable course to follow.

There are other situations in which physicians are confronted with impossibilities where a desirable course of action is not feasible, e.g., where it is impossible for some reason or other to save a person's life. Some impossibilities are natural and some, of course, are social, institutional and legal. Physicians, like the rest of us, often feel that they ought to do something for their patient, but bureaucratic regulations or other barriers make it impossible. Sometimes, again, it is impossible for a physician to do what the internal morality of medicine requires him to do for a patient, because the patient will not comply. He is a 'difficult' or 'ugly' patient. Physicians are so used to being in a position of power where they are able to do what they think is right (according to a medical norm) that it often comes as a shock to them to discover that there are many things that are impossible, things that ought to be done but that cannot be done – for practical social and psychological reasons and sometimes even for moral reasons. I shall return to the maxim, 'ought implies can', later in another connection.

Finally, a brief commentary is necessary concerning the relationship between the norms of the internal and external moralities of medicine that we are examining here and decision-making. In my discussion of the internal morality of medicine I shall deliberately bracket the question of who should make a particular clinical decision and how it should be made, e.g., whether it should be made unilaterally by someone in authority or collectively by consensus. What I am concerned with in this essay is what might be called 'input', that is, the kinds of moral considerations, norms and arguments, that ought to be taken into account by a decision-maker, whoever that might be, and in the decision-making process, whatever form it may take. As I have already suggested, I am interested here in the *what* (or *why*) of a decision and not its *who* or *how*! [8]

THE MORAL PROFESSIONS IN GENERAL [9]

The easiest way to understand what the internal morality of medicine is all about is to begin with some characteristic traits of professional morality in general. For it will soon be obvious that professionalism provides the background and essential framework for the ethics of medical professionalism, that is, the internal morality of medicine.

Medicine is one of the so-called 'higher' or 'learned' professions, which include law, ministry and the scholarly professions. Because these particular professions typically claim to have some sort of moral foundation and moral dispensation and their members often play the role of moral entrepreneurs, I choose to call them the *moral professions*. By using this label, I can distinguish the professions I wish to discuss from other professions that have no similar moral pretensions, such as architecture or computer programming. Let us examine some other typical features of the moral professions.[10]

First, in addition to the usual marks of a profession, such as requiring advanced, specialized training and licensure or accreditation, the moral professions are distinguished from many other professions in that they are 'service oriented', that is, they provide services of a special sort to individuals, variously called 'clients', 'patients', 'parishioners', or 'students'. Thus, the moral professions typically, or in most instances, involve some sort of personal relationship between the professional and those whom he serves. In addition, the service provided by a professional to the individual served is said to be 'altruistic', that is, it is intended to be for the benefit of that individual. A professional is supposed to 'do everything possible' for his clients, patients, etc. He acts *zealously* in their behalf. Furthermore, the professional is not supposed merely to please his clients, patients, etc., but to do what he thinks is good for them and to serve their needs rather than just their wants. In this respect, the services provided by one of the moral professions is quite different from the service that a person receives from, say, a merchant or a purveyor of services such as a taxi-cab driver. The latter are supposed to comply with the customer's bidding – for a price – and are not supposed to inquire into what is good for him. Moreover, their motive can, quite licitly and respectably, be to make a profit – to make more money by selling wares and services to customers. In our bourgeois social system, it is eminently proper for a merchant to put his own interest first and what benefits the customer counts only in relation to that interest.

Lawyers and physicians are, of course, also interested in money, but making money is not supposed to be their primary objective. What they do for a client or patient is not supposed to be determined by monetary considerations as it is, quite properly, with a merchant. The point is not an empirical one, nor a psychological or sociological one, rather it is a logical point or, if you wish, a conceptual truth. For, although one would not be surprised to hear a merchant say that he sells shoes to make money, it would be surprising and highly improper for a physician to say that he performs mastectomies to make money. He would say, or at least is supposed to say, that he performs operations because his patients need them. In this regard, the aims of the moral professions are more like those of other public service occupations, such as the police or librarians, whose aims are to serve others rather than to make money. (The police are not supposed to make arrests in order to make money!)

An important characteristic of the moral professions is their claim to professional autonomy, that is, to self-governance. Each of the moral professions claims control over the services it provides, over its standards of competence, and over its membership. The aims, standards, procedures, and qualifications for membership are set by the profession itself, either formally or informally. The moral professions are, or claim to be, self-policing; they exercise surveillance over the services provided by their members and discipline them for non-compliance with set standards – in theory, if not in practice. According to the principle of autonomy, the evaluation of physicians and of their services is intra-professional. Although in fact the professions are beginning to lose some of the 'collegial' control that they traditionally have possessed, they still claim that only if they are free from outside regulation will they be able to provide the services properly expected of them by their clients and patients. It is clear that, in one way or another, the concept of professional autonomy plays an important, although perhaps not indispensable role, in the internal morality of medicine. In any case, it is part of the background for moralities of this type.

Each of the moral professions builds on what Parsons calls an 'intellectual component', an esoteric body of learning. The services provided by the profession are based on and are legitimized by this body of learning. In the case of medicine, the intellectual component consists chiefly of biomedical science, broadly construed to cover not only biochemistry, but also various branches of clinical medicine. The scientific character of medicine serves to distinguish professional medicine from amateur medicine of various sorts: folk remedies, holistic medicine, magic, faith-healing, and so on. Orthodox medicine, as contrasted with various unorthodox methods, is based on science. Its scientific base is used to justify the physician's claim to special consideration, to special rights and privileges, and to monopolistic control over the delivery of health care. In turn, it provides the rationale and guidance for medical practice, procedures, and protocols. There is a strong feeling, widely accepted in our society as well as in medical circles, that to opt for medical procedures that are not based on medical science is to yield to mysticism, superstition and other forms of irrationalism. Science may not be all of medicine, but surely it is an essential ingredient in medicine as it exists today. Without science there is no medicine, although, as we shall see, this is only half of the story.

Each of the moral professions has what will be called a *doctrine*. It could equally well be called the 'ideology of the profession'. The doctrine is a set of teachings about the aims of the profession, the kinds of services and benefits that it provides for its clients and patients, the procedures that should be followed in pursuing its aims, and so forth. For our purposes, it is important to note that the doctrine defines, either explicitly or implicitly, what is good for the client or patient and, accordingly, the nature of the benefit that professional service is supposed to bestow. In the case of medicine, this part of the doctrine relates to what are called the 'goals of medicine'. It will become clear as we proceed, that there is a great deal of confusion concerning precisely what these goals are and how they are related to medical practice. In view of the intimate logical connection between the goals of medicine and the internal morality of medicine, we will need to examine and appraise critically various conceptions of these goals. For obvious reasons, I cannot offer a definitive answer to the question: what are the goals of medicine? But I shall try to show some directions that our search for an answer might take. In the meantime, I want to warn against the simple-minded assumption that the question is a purely scientific one. The conception of the goals of medicine may, indeed, be grounded in science and have some scientific content, but it must also be a value-concept of some sort or other. For, despite asseverations to the contrary, science and value cannot be separated in medicine: medicine is and must be a 'value-laden science'.

ESOTERIC SERVICES AND THE GOALS OF MEDICINE

The keynote of this essay is the proposition that the moral professions provide esoteric services for the benefit of their clients. It follows that the basic aim of medicine is to do what is best for the patient. I do not wish to claim that medicine does not or ought not concern itself with other things, e.g., public health, but, at least as far as clinical medicine and the physician-patient relationship are concerned, the central focus is on the patient and what is good for him, what he needs.[11]

What, then, are these esoteric services that are provided by medicine? What kind of benefit do they bring to the patient? What, more specifically, are the aims of medicine? In answering these questions the obvious tool of analysis that we must use will be the means-end category; for all of the questions we need to ask are questions about ends and means.[12] What is the purpose of a procedure, an operation? Are the means, e.g., the prescribed treatment, the most effective and acceptable means to the end? What is the end? How does it fit in with other values? And so on. A critical analysis must ask such questions both with regard to general practices, e.g., the use of exotic technology, and with regard to decisions in particular cases. Without intending to deprecate the efforts of individual physicians and the thoughtfulness with which they approach these questions, I want to call attention to some questions that need to be asked in connection with the aims of medicine.

We may start with two standard answers to the question: what is the aim of medicine? Both answers are, I shall urge, unsatisfactory, but for opposite reasons. The first is the classical answer going back to Plato and Aristotle, and reaffirmed by Leon Kass, that the aim of medicine is health.[13] The second answer, also of ancient vintage, is that the aim of medicine is the cure and prevention of disease. Where the elimination of a disease is not possible, the aim would be the alleviation of disease; I shall call it the reduction of disease.

It should be recognized at the outset that both 'health' and 'disease' are normative or value concepts.[14] Health is something that is by definition desirable and disease is something that is by definition undesirable. Sometimes one of these concepts is reduced to the other; it is defined as the privation of the other. Thus, 'health' is defined as the absence of disease or 'disease' is defined as the absence of health. For reasons that I have presented elsewhere, the proposed reduction is invalid.[15] Therefore each of the concepts, health and disease, has to stand on its own two feet, as it were, and each will be seen to generate an entirely different definition of the aim of medicine.

Without attempting to define 'health' here, let us assume that it refers to a positive, general condition of some kind, a state of somatic, perhaps also mental, well-being, of the sort that we have in mind when we say of an individual that he is in good health and of another one that he is in poor health. There are several obvious objections to using health in this sense to define the aim of medicine. I can only describe them cursorily. First, the etymology of 'health' suggests that health is holistic and that it applies to the organism as a whole. That means that any undesirable condition, such as athlete's foot or a wart, is so because of its relation to the condition of the body as a whole. The generality of the concept does not fit well with the specificity of the services provided by the physician. Again, the concept of health is vague, that is, it tends to merge with other concepts of well-being, such as happiness, welfare, soundness, adaptability, and so on. But it should be clear that it is not the province of a physician to concern himself with every aspect of a person's well-being, although it is sometimes held that physicians should try to solve their patients' psychological and social problems as well as their somatic ones. Finally, if we include under health resistance to disease, the ability to throw off and to recover from disease, then there are many non-medical factors that are as important, if not more important, than medical ones in attaining and retaining health. It has been quite conclusively shown, for example, that nutrition is more important for health than, say, medical care. Indeed, poverty is a greater threat to health than the unavailability of medical services. Personal habits – life-style – are also more important determinants of good health than visiting a physician.[16]

Considerations such as these suggest that, although good health is an unquestioned benefit for everyone, it is by no means a goal that is the monopoly of physicians; indeed, the esoteric services provided by physicians have a minimal impact on health in general. For these reasons, the health conception of the goal of medicine is not very useful for identifying the kind of esoteric services to be provided by medicine. Furthermore, it is too general and vague to provide an answer to the question: what is the point of having such and such medical treatment? To say: "Because it is good for your health," may be true, but it is hardly an adequate answer.

Turning to the reduction of disease definition of the goal of medicine, we encounter precisely the opposite problem, namely, instead of being too general, this definition is too narrow and specific. Again, this is no place to discuss the definition of 'disease', but the most useful notion of disease refers it to the category of specific identifiable diseases that are sought as the end-product of the diagnostic inquiry. The list of diseases is very long. It covers a wide array of lesions, disorders, derangements, dysfunctions and abnormalities. As Wulff writes: "The clinician today must recognize that the present disease taxonomy is arbitrary, imperfect and ever-changing, and at the same time he must realize that we cannot do without it" ([23], p. 66). For our purposes, we may regard 'disease' as referring to such things as diabetes, cancer, tuberculosis, myocardial infarction, and so on. It implies that there is something amiss.

There is, of course, some point to saying that the goal of medicine is to cure or reduce disease in this sense, that is, specific diseases. For, after all, clinical medicine consists largely of the diagnosis, prognosis and treatment of specific diseases, almost all of which are listed in a standard textbook of internal medicine.

However, the question arises: why should disease be treated? More specifically, we may ask whether it is the goal of medicine to treat each and every disease that is identifiable by the clinician? Some diseases on the list are insubstantial; some are hardly noticeable, such as being a carrier for a rare genetic disease. Is it not the function of a physician to identify and treat serious diseases, that is, diseases that are life-threatening, disabling, debilitating or painful? Clinical judgment requires discrimination and selection among diseases and, in particular, among the diseases that he discovers in his patient. The physician's job may be to reduce disease, but only selectively. Again, we must be careful with other uses of the term 'disease', because we often use the term for any condition that calls for medical treatment, i.e., treatment by a physician. Thus, alcoholism is now regarded as a disease, and homosexuality is no longer regarded as a disease. So, in one way or another, the concept of disease is manipulated to make it conform to a previously accepted conception of the goal of medicine rather than the other way around.

In any case, the disease approach to the goal of medicine has a number of drawbacks, for it is either quite indiscriminate, or else it involves us in a circle that throws doubt on its usefulness for our stated purpose, namely, to determine what precisely is the aim of the esoteric services provided by medicine. That is not to deny that one of the things that physicians do is to treat diseases.

Finally, we face another, perhaps more serious, problem if we take the aim of medicine to be to identify and treat disease, to cure, to mitigate or to stave off disease. For what are we, then, to say about chronic, incurable and intractable diseases such as multiple sclerosis or terminal stage cancer? Since these conditions are conditions that the physician can do nothing to reduce, must we conclude that they are outside the province of medicine? Do they fail to meet the definition in the same way that vasectomies and non-medical abortions fail to meet Kass's definition? Is it not the job of a physician to do something for a patient whose disease he cannot cure? The disease definition would seem to imply otherwise.

At this point, some comment is needed on the kind of definition that we are seeking and how it is to be used. For reasons that go deeply into philosophy of language, we cannot and should not look for a definition setting forth the necessary and sufficient conditions for taking a physician's service (or non-service) to be medical or not or for a goal to be medical or not. The subject-matter of our present inquiry is not susceptible to rigorous definitions of this kind.[17] Hence, there is something logically objectionable in trying to deduce Dos and Don'ts directly from a concept of the aims of medicine as Kass tries to do. Rather, our search here is for what may be called a *focussing* definition, that is, it is a definition that directs our attention to what is morally central and away from what is morally peripheral in an activity (or objective) that is being scrutinized. By focussing our attention, the 'definition' says, in effect, this is where to look in understanding and appraising what is morally important in what physicians do in their capacity as physicians, that is, as members of the medical profession. When we have done this, then we should be in a position to grasp the (correct) principles of the internal morality of medicine.

THE PROBLEMS APPROACH

Since neither of the two traditional answers to the question concerning the aims of medicine appears to be satisfactory, it might be better to try a quite different approach. The one I suggest might be called the 'problems approach'. The problems approach is more down to earth than the other approaches and corresponds more closely to what is generally expected from moral professionals and to what they, in turn, generally feel obligated to provide. Inasmuch as there are many facets to the problems we shall be concerned with and they are also very complicated, my discussion will of necessity be tentative and speculative; the whole idea of a problems approach along the lines I suggest needs to be worked out in much greater detail.[18]

The problems approach is based on the proposition that it is the function of a professional, particularly of someone from a moral profession, to help the client or patient solve a problem or a set of problems that he faces or is likely to face. Thus, a patient seeks a physician's help when he has a problem, e.g., when he is not feeling well. The physician's job, in turn, is to help identify the problem, to recommend a solution, or even to bring about a solution through some action of his. The patient (or client) seeks help because he is, in a certain regard, 'helpless'; the professional is able to help because of his special expertise, based on training and knowledge, and his skill, all of which amounts to saying that he is 'learned' in medicine.

Unlike the other approaches already examined, the problems approach is multi-dimensional. Because each problem under consideration has many different dimensions, there are many ways of analyzing it and many different questions that can be asked about it. In other words, problems have what might be called a 'deep structure'. In contrast to the problems approach, the two approaches already discussed, the health and disease views of the aims of medicine, might be described as linear, since they analyze the issues along a single line connecting means and ends. As a result, their conclusions are inevitably simple-minded and superficial.

There are a number of advantages of the problems approach that might be mentioned right away. First, it explains why we go to a physician rather than to a gym-teacher or a dietitian when we have a 'health problem'. And it explains why we do not go to a physician for the treatment of any and every disease that we have; we go only when the disease has become a problem for which we need help. Non-problematic diseases, e.g., a common cold, do not require the services of a physician. Second, unlike the other approaches, the problems approach is concrete and particular, rather than general or abstract. It is therefore easier to apply to specific issues in the internal morality of medicine. Finally, as I shall try to show, the problems approach provides a guide for overcoming differences in outlook between physicians and patients concerning what ought to be done in a particular medical situation.

THREE ASPECTS OF A PRACTICAL PROBLEM

The problems that we are concerned with in this essay are *practical* problems. They are in important ways quite different from theoretical problems, such as those that mathematicians grapple with. A practical problem is a question about what to do about something. The answer to the question tells us what to do and the doing of it leads to the successful solution of the problem. Following Dewey, we can identify three stages in the process of problem-solving: first, the perception of the problem; second, the investigative or deliberative process; and, third, the realization of the solution.[19]

There are three sides to a practical problem, roughly corresponding to these three stages, that are particularly useful for understanding and dealing with particular practical problems. A satisfactory identification and analysis of any particular problem requires answers to questions about each of them. Each of the three is individually necessary and they are jointly sufficient for a question to be also a practical problem. (Of course, due provision must always be made for the 'fuzzy' and open-textured character of any practical concept in what I shall say about practical problems.)[20]

The *first* side of a practical problem is that it refers to an evil, an undesirable condition of some sort or other, to be gotten rid of or to be mitigated.[21] A problem exists only when something is wrong. When everything is all right, there are no problems. The poor have problems, the rich do not. (I am speaking, of course, of financial problems). The sick have problems, the healthy do not. When a person says: "I have a problem," he implies that something is wrong that needs to be corrected and not simply that he is curious about something (theoretically). This aspect of practical problems explains why we use the possessive about problems: "That's his problem, not mine." For evils themselves are normally referred to particular persons, whose evil or undesirable condition is in question, or, are at least, they are somehow associated with persons who are concerned about them. So, the possessive indicates which persons are involved in the attribution of a problem. Thus, in general, practical problems are associated with individuals, whose problem they are. There is no such thing as an 'impersonal' problem, although, of course, there are social problems, group problems, and problems for everybody, such as the threat of nuclear war. Thus, it is always appropriate to ask someone who alleges that there is a problem about *X*: "What is wrong?" "What is the matter?" "What is the evil?" And, ordinarily, it is also appropriate to ask: "Whose problem is it?" or "For whom is it a problem?"[22]

The *second* side of a practical problem is that it implies a question and reflects a perplexity, an uncertainty. To say that Jones has a problem, say, a financial or a health problem, means that he or others concerned with his welfare are uncertain about what to do. Sometimes problems arise out of dilemmas where it is unclear which of two incompatible courses of action one ought to follow: one is uncertain which to do. There are, of course, many other sources of uncertainty that lead to practical problems.

In the absence of an uncertainty, a problem simply becomes a task, that is, something to do rather than something to ask questions about. Thus, for example, washing dishes is not a problem unless there is some question about who is to do it or when it should be done. Many evils that we have to deal with do not generate problems, for we already know what to do about them. They simply create tasks.

The fact that problems (of the type we are considering) imply questions, perplexities, uncertainties, means that it always makes sense to ask of a person who says he has a problem: "What is it that you don't know?" "Why are you uncertain?" "Why is it a question?"

Third and finally, a problem calls for a solution, ordinarily a doing or a not-doing something; for, as I have already said, a practical problem is a question about what to do. The solution is the elimination or mitigation of the evil by something we can do. If it is perfectly clear that there is nothing that can be done about an evil, then there is no problem. An earthquake, getting older, or even death may lead to problems, but they are not themselves problems. Past events are not problems. The holocaust is not a problem for us today, because its being in the past we can do nothing about it. That is not to deny that the consequences of the holocaust are still with us and create problems for us today. As Aristotle said, we do not deliberate about the past; we can regret the past, but cannot change it. Problems, however, are the kind of things that we do deliberate about. By the same token, if a person is dying of incurable cancer, the cancer is not itself a problem. Whatever problems there are are things connected with the cancer that one can do something about; for example, one can relieve the victim's suffering, his anxieties, and his grief, and one can help others to cope, and so on.

THE IDENTIFICATION OF PROBLEMS

In order to understand, analyze and cope with a particular practical problem, the first task is to identify the problem, that is, to determine exactly what the problem is. Because problems have, as we have seen, many facets and are frequently very complicated, the identification of a problem is not always an easy task. Unfortunately, professional problem-solvers, such as clinicians, often do not take the question of identification very seriously. The three sides of a practical problem that I have pointed out provide us with guidelines for identifying a problem: they give us three sets of questions to be asked concerning any problem. (a) What is the evil or undesirable condition that creates the problem? (b) Where is the uncertainty that makes it into a problem? (c) Is it possible to do anything about (b) and, if so, what kinds of action or non-action might lead to a solution of the problem, i.e., the elimination or mitigation of (a)?

A satisfactory identification of a problem requires suitable answers to *all* of these questions. Inasmuch as the three facets of a problem are logically interconnected, in order to be suitable an answer must be relevant in the sense that its connections with the other answers are perspicuous; for example, the action to be taken must be related to the evil to be overcome.

Underlying the view of problems adopted here is the assumption that it is always possible to be mistaken about the problem in that one may not have the right answer to some of the questions just mentioned. Accordingly, one may misconceive or misconstrue a problem and one may see a problem where there is none or not see a problem where there is one. For example, a person may go to a physician thinking that he has cancer (i.e., an evil), but it may turn out after the physician examines him that he really does not have cancer at all (i.e., no evil). If he has a problem, it might be a quite different kind of problem, e.g., a psychological problem. In that case, we could say that the patient has misidentified the problem. Not only patients, but also clinicians may misidentify problems.

Problems can be simple or they can be complicated, they can be limited or they can be unlimited. (An unlimited problem is one that is capable of a number of solutions.)[23] A simple problem is limited in that one knows what is wrong, knows why one is uncertain, and knows what the solution would be. Consider the problem of finding a lost key: one knows what is wrong – the key is missing; one knows what is uncertain – its whereabouts; and one knows what to do, i.e., to develop a plan of action or strategy for looking for it.

Many clinical problems, or are at least assumed to be, like the problem of the lost key. A patient has a sore throat or a rash. The physician makes a diagnosis and prescribes a treatment, say, an antibiotic, and the sore throat or rash disapprars. The physician has effected a cure, which is the solution to a particular kind of practical medical problem. At the primary care level, such problems are normal. But as cases become more complex, it becomes increasingly difficult as well as more urgent to identify the problem and so to determine what should be done.

A PARADOX

At this point, I should like to call attention to an apparent paradox in clinical medicine relating to the identification of the problem in a clinical case. How do we determine what the evil or undesirable condition is that defines the problem and stipulates the solution? Is the evil or undesirable condition in question the patient's original complaint, e.g., abdominal pains, sore throat, or weakness, that is, the so called iatrotropic symptoms? Or, on the other hand, is it the disease as identified by the physician that explains the patient's symptoms and that needs to be treated? Sometimes, indeed, the disease discovered by the physician is unrelated to any iatrotropic symptoms. What, in such cases, is the problem? The practical paradox involved here is described by Wulff as follows:

> Iatrotropic symptoms are very important as they usually represent that problem which the doctor, in the eyes of the patient, has to solve. . . . Nowadays it is not rare in hospital practice that investigations bring some unexpected results to light and these lead to more examinations along a side tract. After a while the whole staff is interested in, say, the immunoglobulin pattern, and nobody remembers why the patient was admitted. Only on the day on which the patient is discharged will he say, 'But you have not done anything about my backache!' ([23], p. 10).[24]

The paradox of having a problem that is different for patient and for physician creates not only the practical difficulty of reconciling the interests of patient and physician, but also a deeper philosophical quandary concerning what kind of problem it is that furnishes the rationale for the patient-physician transaction and that, more generally, defines the goal of medicine.

The paradox is reflected in different ways of conceiving the problem, how it unfolds, and how it is solved. The patient thinks of the iatrotropic symptom (complaint) as caused, e.g., by streptococcal infection, and thinks of the problem as solved by the treatment that cures his strep throat. The disease is regarded as a controllable cause, and the symptom and the cure as effects. The physician, on the other hand, takes the symptoms as *manifestations*, which also include laboratory findings, and uses them as evidence or clues for the identification of the disease. Each mode of conception, reflected in the language used, begs the question as to which is basic and which is derivative (dependent, secondary).

DIALECTICAL REASONING AND THE SOLUTION OF THE PARADOX

What is the solution to the paradox? Must we adopt a relativistic attitude and say that problems are relative to the perceivers and that, consequently, there are two distinct problems, the patient's problem and the physician's problem? Or, on the other hand, must we opt for one of the alternatives and assume either that the real problem is the patient's problem (his complaint) or that the real problem is the physician's problem (the disease)?

For a number of reasons, all three suggestions are unacceptable, although admittedly there are many authors who have adopted one or the other of them. They are objectionable on a number of counts, theoretical and practical.

To begin with, if we assume that the general aim of a moral profession such as medicine is to *help* people with their problems, then we must either reject that view of the moral professions or else reject all three suggestions. For '*X* helps *Y* with regard to a problem *P*' entails that '*P*' means the same thing for *X* and *Y*. *X*, the patient, and *Y*, the physician, are concerned with the same problem, *X* seeking the physician's help for the problem and *Y*, the physician, helping *X*, the patient, with the same problem. Thus, if we want to preserve the notion of helping as an essential ingredient in medical professionalism, we must show how it is possible for both parties, patient and physician, to be concerned with the same problem.

Now, unlike the views I am arguing against, e.g., relativism, the position defended here presupposes that it is possible for a person to be mistaken in his identification of a problem: It is both theoretically and practically possible to misidentify it, to misconceive it and to misconstrue it. Hence, it is perfectly possible in cases where there appears to be a paradox that *both* parties might be wrong about the problem, i.e., that the problem is not one exclusively centered on the patient's complaint or one exclusively centered on the physician's diagnosis of a disease. Somehow the two positions must be brought together. But a necessary condition of such a reconciliation is that neither point of view be categorically rejected for the sake of saving the other. To speak metaphorically, they must be regarded as two aspects of a single whole, the whole being a complex of discomfort, disability, disorder, disease, disquiet, etc. Exactly how these elements are connected and combined is a complicated matter that cannot be dealt with here. The key to the synthesizing process will be found in questions about the three facets of a problem, particularly, in questions about what evil or undesirable condition is implied by the problem and what makes it evil or undesirable. If we take, as an example, a disease like acute bacterial meningitis, we can easily conceive how this kind of question about why it is undesirable could be answered.

Reasoning of the kind required here may be said to be 'dialectical', for it is a logical process that proceeds by developing questions and answers in particular contexts to meet certain needs and with particular objectives in mind. Pre-eminently it is a social process, which is sometimes called 'dialogue'. We can see that in fact we employ dialectical procedures all the time in dealing with practical problems of a social nature. The paradox that we began this discussion with may in fact be more conjectural than real. After discussion and reflection, a patient is usually quite ready to change his view of what his problem is from purely a symptomatic one to one that ties the symptoms, present and possible, to a disease, and the physician, for his part, is often quite ready to adapt his notion of a problem, diagnosed in terms of a particular disease, to what it will mean to the patient in the short run and in the long run.

CONCLUSION AND SUMMARY

Further comments on the identification of problems and on the problem of how to reconcile different conceptions of a problem would require a more extended analysis of the logic of problem solving and of clinical judgment and decision-making. My principal aim in this essay has been to suggest a new approach to the ancient question of the aim of medicine and to indicate some directions that it might take. Underlying the suggestions made here is the assumption that we are dealing with very complicated matters and that no simple solution is possible. With this in mind, I propose that by concentrating on problems, the more specific the better, we may be able to pinpoint the nature of the esoteric services that a profession like medicine provides to patients, to clear up misunderstandings about what these services are and should be, and to recognize the limits of medicine that are set by the limits of the ability of physicians to deal with the problems that are brought to it. Finally, the problems approach may help to define more adequately than the other approaches what is proper and what is improper for a physician to do to and for his patients. But it should be borne in mind that the ideas I have presented here are highly speculative and should be taken as suggestions only. Many details need to be worked out and important lines of thought need to be developed before my proposals could in any sense claim to be definitive.[25]

Brown University,
Providence, Rhode Island

NOTES

[1] I shall use the term 'norms' as a generic concept standing for authoritative rules, principles, ideals, goals and values, without regard to whether they are or ought to be accepted. The term the 'internal morality of medicine' is adapted from Lon L. Fuller, who introduced the terms 'internal' and 'external morality of law'. My use of the terms has some affinities to Fuller's use, but is in some respects quite dissimilar. See [7].
[2] See [18] and [21].
[3] For a philosophical exposition of this point about rules and concepts, see [11].
[4] The norms in question are accepted norms. Whether or not they are acceptable, that is, warrant being accepted, is a separate issue. An account of the importance of these norms for surgical practice may be found in [2].
[5] These phrases are taken from ([20], p. 641).
[6] See Note 1.
[7] The term 'prima facie duties' comes from [19]. My own views on the subject are set forth in [12].
[8] For my views on these other topics, see [14].
[9] In this section I draw on another essay [15].
[10] In this paragraph and the following ones, I draw heavily on the sociological literature on the professions, in particular [8] and [9].
[11] Many of the professions have official gatekeeping jobs that furnish certification for a particular status or position. In our society, a physician's certification is necessary for birth and for death.
[12] It should be noted that in this regard the internal morality of medicine differs from the external morality of medicine, for the latter uses other ethical categories besides the means-end category, e.g., the category of rights and responsibilities, moral rules, and moral virtues.
[13] See [10].
[14] For a collection of articles on this subject, see [3].
[15] My own position on the concepts of health and disease is set forth in [13].
[16] See [16].
[17] Critiques of rigorous definitions of this kind are common in contemporary philosophical literature. They reflect the influence of L. Wittgenstein [22].
[18] The notion of a problems approach is reminiscent of what has been called the 'problem oriented approach in medicine (POMS)'. There are similarities and differences between these two approaches. The emphasis in POMS is on the problem oriented medical record, which is used to formulate problems, assessments, diagnoses, etc. My own approach, as will become clear, goes beyond the POMS approach, which, in my mind, underestimates the theoretical as well as practical difficulties in identifying problems. On the whole, the POMS approach comes very close to adopting what I would call the patient's perception of the problems. I am indebted to Sarah S. Bishop for calling my attention to the POMS. See [1].
[19] The discussion that follows is greatly influenced by the writings of John Dewey. See, in particular, [6].
[20] The notion of fuzzy and open-textured concepts is due to Wittgenstein.
[21] This aspect of problems is explained very clearly by John Passmore in [17].

[22] The problem of ownership of problems is slightly more complicated than this, because we want to say that one person can 'unload' his problems on another or that another person can take on one's problems as an *alter ego*. This is precisely what happens when one person helps another. See below.
[23] "An unlimited problem is that which has innumerable answers." Hutton, 1798, as quoted in ([4], p. 2310).
[24] Iatrotropic symptoms "are those symptoms which made the patient turn to his doctor" ([23], p. 9).
[25] Earlier drafts of this essay were read to a faculty group at the Harvard Medical School and to a seminar on medical ethics at Union College. I am indebted to my audience and friends at both places for their helpful comments and criticisms. The ideas in this essay were originally developed in a course in philosophy of medicine that I gave at Brown University with my friend and colleague, Robert P. Davis, M. D. I am particularly indebted to Dr. Davis and to our students for many stimulating and informative discussions on the aims of medicine.

BIBLIOGRAPHY

1. Bishop, S. S.: 1980, 'Explanations in Medicine: The Problem Oriented Approach', *Journal of Medicine and Philosophy* 5 (March), 30–56.
2. Bosk, C.: 1979, *Forgive and Remember*, University of Chicago Press, Chicago.
3. Caplan, A. L. *et al.*: 1981, *Concepts of Health and Disease*, Addison-Wesley Publishing Co., Reading, Mass.
4. *The Compact Edition of the Oxford English Dictionary*, Vol. II: 1971, Oxford University Press, New York.
5. Curran, W. J.: 1980, 'The Ethics of Medical Participation in Capital Punishment by Intravenous Drug Injections', *New England Journal of Medicine* 302 (January 24), 326–330.
6. Dewey, J.: 1957, *Human Nature and Conduct*, Modern Library, New York.
7. Fuller, L. L.: 1969, *The Morality of Law*, Yale University Press, New Haven.
8. Goode, W. J.: 1969, 'The Theoretical Limits of Professionalization', in A. Etzioni (ed.), *The Semi-Professions and Their Organization*, Free Press, New York, pp. 266–313.
9. Hughes, E. C.: 1965, 'Professions', in K. S. Lynn (ed.), *The Professions in America*, Houghton Mifflin, Boston, pp. 1–14.
10. Kass, L. R.: 1981, 'Regarding the End of Medicine and the Pursuit of Health', in A. L. Caplan *et al.* (eds.), *Concepts of Health and Disease*, Addison-Wesley Publishing Co., Reading, Mass., pp. 3–40.
11. Kovesi, J.: 1967, *Moral Notions*, Humanities Press, New York.
12. Ladd, J.: 1958, 'Remarks on the Conflict of Obligations', *Journal of Philosophy* 55, 811–819.
13. Ladd, J.: 1982, 'Concepts of Health and Disease and Their Ethical Implications', in K. Nelson and B. Gruzalski (eds.), *Value Conflicts in Health Care Delivery*, Ballinger Press, Boston.
14. Ladd, J.: 1982, 'The Distinction Between Rights and Responsibilities: A Defense',

Linacre Quarterly 49 (May), 121–142.
15. Ladd, J.: forthcoming, 'Philosophy and The Moral Professions', in J. Swazey and R. Fox (eds.), *Social Controls and the Professions*.
16. McKeown, T.: 1979, *The Role of Medicine*, Princeton University Press, Princeton.
17. Passmore, J.: 1974, *Man's Responsibility for Nature*, Scribners, New York.
18. Rachels, J.: 1979, 'Euthanasia, Killing and Letting Die', in J. Ladd (ed.), *Ethical Issues Relating to Life and Death*, Oxford University Press, New York, pp. 146–163.
19. Ross, W. D.: 1930, *The Right and the Good*, Clarendon Press, Oxford.
20. Siegler, M.: 1981, 'The Doctor-Patient Encounter and Its Relationship to Theories of Health and Disease', in A. L. Caplan *et al.* (eds.), *Concepts of Health and Disease*, Addison-Wesley Publishing Co., Reading, Mass., pp. 627–644.
21. Tooley, M.: 1979, 'Decisions to Terminate Life and the Concept of Person', in J. Ladd (ed.), *Ethical Issues Relating to Life and Death*, Oxford University Press, New York, pp. 62–93.
22. Wittgenstein, L.: 1953, *Philosophical Investigations*, trans. by G. E. M. Anscombe, MacMillan, New York.
23. Wulff, H.: 1981, *Rational Diagnosis and Treatment*, 2nd. edition, Blackwell Scientific Publications, Oxford.

GEORGE J. AGICH

SCOPE OF THE THERAPEUTIC RELATIONSHIP

The goal of this essay is to discuss the scope of the therapeutic relationship primarily as a conceptual question. Because major intrinsic and extrinsic changes in medicine have complicated considerably relationships between physicians and patients, it is exceedingly difficult even to speak of the physician-patient relationship without engaging in ideological rhetoric [34]. Conflicts involving, for example, informed consent and the common law notion of 'therapeutic privilege' and confidentiality and the duty to warn others of potentially violent actions on the part of patients raise fundamental questions regarding the limits of therapeutic relationships.[1] Although the notion of 'scope' may be construed as indicating the drawing of boundaries simply in terms of the concepts of rights and duties, it is important to recognize that the various rights and obligations which constitute therapeutic relationships and physician-patient relationships in particular do so in terms of a specific understanding of those relationships. This should not be surprising insofar as the meaning of rights (and duties) is functionally related to particular contexts or practices. A full analysis of the scope of the therapeutic relationship, then, will necessarily indicate how rights and duties are woven into the concept of the relationship within the practice of medicine. But before such an analysis can be undertaken, it is essential to clarify the meaning of medicine as a practice and the definition of therapeutic relationships in medicine. It will be useful also to consider two ways that the question of the scope of the therapeutic relationship has been raised: in terms of models of physician-patient interactions and the distinction between research and therapy.

MEDICINE AS A PRACTICE

Medicine is a rich and extensive practice which makes analysis of the therapeutic relationship a rather arduous undertaking.[2] Two general aspects of practices, however, make such an undertaking less formidable. The first involves the rule-governed character of practices. A practice is a set of ordered activities characterized by patterns of regularities. The term 'rules' refers not only to the seemingly 'objective' regularities observed in the behaviors of

Earl E. Shelp (ed.), The Clinical Encounter, 233–250.

participants in a practice and catalogued by social scientists, but also to the regularities and norms which define the practice *for the participants* and which participants in the practice knowingly and intentionally follow. Rules in this sense are less like laws of nature, objective explanatory frameworks, than they are like rules of thumb or recipes for action which all social actors, including social scientists, employ in engaging in particular practices. So regarded, 'rules' refer to the specialized knowledge, skills at hand, and normative procedures which define the very status of individuals as practitioners or participants in the practice in question. In other words, participants in a practice knowingly – though not necessarily self-consciously – follow rules which define the practice. Thus, following rules is one defining feature of the status of participants in a practice.

Rules are not free-standing, though; they depend on a logically distinct, but related, set of values and beliefs which provide participants in a practice with a basis for accepting and following rules. Rules serve to guide conduct only where there is acceptance of ways of thinking and acting which have a wider significance and relevance than the rules themselves. Values and beliefs provide the basis for participants to regard rules as valuable and as compelling their obedience. Logically, this understanding cannot be part of the formulation of any given rule or set of rules ([12], p. 106). Beliefs and values thus comprise the second important feature of practices.

Most disputes regarding the scope of the therapeutic relationship are naturally expressible in terms of rights and duties and implicitly refer to rules which define the practice of medicine – for example, cases of invasion of privacy and unauthorized treatment are expressed in terms of patient autonomy and informed consent. But because problems arising in the practice are often articulated by participants only at the logically distinct level of rules – and then often only in terms of authority, power, and rights – values and beliefs often go unexamined. Although unexamined, tacit disagreements at this level contribute significantly to disputes and conflicts regarding rules. For this reason, it is essential that the values and beliefs defining therapeutic relationships in medicine be made explicit. That is not, however, an easy task.

Because therapy can be a piecemeal affair or 'team' effort, isolating the therapeutic relationship and delimiting its scope is difficult. Although the physician-patient relationship is generally acknowledged as the primary therapeutic relationship in medicine – a primacy due in large part to the power associated with the physician's role – it is not a simple or single relationship as the phenomena of referral and consultation attest [36]. Diagnostic and therapeutic activities are intertwined with administrative and bureaucratic functions to achieve the complex goals of institutionalized medical practice.[3] As a result, there are many relationships in medicine which can be broadly termed *therapeutic* (or at least therapeutic in an adjunctive way) beyond that of the physician-patient relationship: relationships with nurses, chaplains, social workers, psychologists, dieticians, occupational therapists, physical therapists, and mental health practitioners. Also, special roles have recently evolved for counselors in death and dying or in grief and mourning and even for philosophers performing 'medical ethics consultations'. In light of this myriad of care-giving roles, the notion of the therapeutic relationship itself becomes problematic. One consequence of acknowledging this array is that the scope of the therapeutic relationship may not admit precise definition, because the therapeutic relationship may not be a unitary phenomenon. Rather, there may be family resemblances between various kinds of therapeutic activities in medicine – and even in physician-patient relationships – which allow overlap and so preclude rigid demarcation. If so, the question of the scope of the therapeutic relationship may necessitate a piecemeal analysis of particular clinical settings in order to clarify the nature of therapy within specific contexts of practice.

Just such an analysis has been undertaken by P. M. Strong in *The Ceremonial Order of the Clinic* [38]. Strong examines the contributions of institutional and bureaucratic structures to the definition of actual therapeutic activities and roles in clinical practice and attends to the special rules which define the actual on-going character of clinical practice – what he calls its 'ceremonial aspect' ([38], p. 6). These rules are clearly influenced by the location of actions within particular clinical or bureaucratic settings. Strong's approach is to emphasize the structural context in which individual actions occur rather than to focus on individuals and their personalities, perceptions, and styles. On his view, rules are indeed present to be followed, but there often are many different ways in which that may be done. In actual practice, one rule may conflict with another. Not only are rules followed in practice settings, but rules are appealed to by participants to justify actions in conflictual interactions ([38], p. 8). Even though Strong's sociological analysis does not deal explicitly with ethical issues, he clearly indicates how various rules actually structure the meaning and limits of therapeutic encounters in particular clinical settings. This approach is important because it shifts the focus of attention from individuals as individuals to the common social meaning and significance of their actions.[4] As a contribution to the phenomenological clarification of the life-world of the practice of medicine, Strong's approach is commendable. But since Strong focuses primarily on the structural and institutional functions of rules, the values and beliefs which make possible the specification of role-formats are mostly unaddressed. To see the inherent limitations of Strong's and the empirical sociological orientation generally, it will be useful to consider more directly the meaning of the therapeutic relationship in medicine from a normative, theoretical perspective.

THE THERAPEUTIC RELATIONSHIP

A therapeutic relationship in general is a relationship between an individual who is seen as suffering from some defect, disability, or discomfort and a practitioner who possesses a specific technical skill or knowledge and occupies a recognized social role. The primary intent of the relationship is to improve the well-being of the suffering individual as the result of action undertaken within the relationship. The definition of the well-being sought and judgments regarding the defect, deficiency, or disability presented are interrelated in such a way as to require reference to the practice of which the therapeutic relationship is a constituent part. Therapeutic relationships thus exhibit three interdependent features: a particular structure (and perhaps style of interaction), a particular intention to improve or contribute to the well-being of an individual suffering a defect, deficiency, or disability (as well as a particular understanding of well-being), and a particular technique or skill possessed and exercised by the practitioner (therapy). Possession of expert knowledge, skill, or technique – in other words, a special power – is thus an essential requirement of being a practitioner. Of course, depending on the therapy and practice in question, it may be dialectical, magical, rational, or ritualistic and more or less formally or rigorously defined.

This definition spells out the essential features of therapeutic relationships generally. Individual types and examples exhibit these features in different ways. One way to indicate the broad range of possible therapeutic relations is to consider generally the kinds of skill, techniques, or interactions that are possible and the correlative definitions of the well-being sought. In these terms not only medicine, but philosophy and art as well have at least historical claim to being therapeutic [7, 37].

Therapeutic relationships may be institutionalized in terms of education requirements, law, or other formal social means – or they may be non-institutional. Medicine in the Greek world appears to have been largely non-institutional. It was a particular craft cultivated by its practitioners without formal or institutional structures and organization. Similarly, folk medicine and healing practices are usually non-institutional. Nonetheless, the relationship between individual clients or patients and practitioners can be regarded as therapeutic insofar as the essential features indicated above are met. These features are determined by socially shared beliefs and values which define the practice in question as a significant social enterprise. If nothing else is to be learned from anthropology, history, and sociology, therapeutic practice must be acknowledged as a universal feature of human experience. So it should not be surprising that a variety of cultural forms might exist. In these terms, the physician-patient relationship must be seen as one of many examples of the therapeutic relationship which must be interpreted in terms of the practice of medicine.

One main requirement of an adequate demarcation of the therapeutic relationship thus involves distinguishing the practice of medicine from other therapeutic practices. If medicine is a unique therapeutic activity or set of activities, then its distinguishing features must be sought not only in terms of the rules which structurally and institutionally define the practice for participants, but more importantly in the social beliefs and values which not only sustain the practice but in the first place contribute to its particular structures and institutional character and complexity. These beliefs and values nowhere are expressed more prominently than in the definitions of well-being and power or knowledge implicit in physician-patient relationships. This point can be developed by considering Philip Rieff's discussion of the striking difference between classical or traditional and modern views of therapy [33].

Modern views of therapy are primarily concerned with 'what cures?' Traditional or classical views are concerned with 'who cures?' ([33], pp. 68–69). On the classical view, it is ultimately the community that cures. The function of the classical therapist is to commit the patient to the symbol system of the community by means of those techniques sanctioned by the community, for example ritual or dialectical, magical or rational: "All such efforts to reintegrate the subject into the community symbol system may be categorized as 'commitment therapies'. Behind shaman and priest, philosopher and physician, stems the great community as the ultimate corrective of personal disorders" ([33], p. 68). According to Rieff, modern views of therapy function where the community itself is disordered and not able to supply a system of symbolic integration. As a result, the therapeutic function of the community is destroyed.[5] On this view, the first requirement for being a therapist of the free individual is to be disconnected from the community: "Not binding sentiment but critical detachment was the attitude most conducive to a sense of well-being" ([33], p. 69).

Although health and disease may be defined on the basis of consensus regarding values and beliefs, cure or successful therapy may not be achievable simply by means of the power of consensus or public opinion.[6] There are obvious limits to the power of consensus to correct or ameliorate defects, disabilities, or discomforts. One cannot expect that even widely shared beliefs, just because they are widely shared, will prevent or cure disease or turn aspirin into birth control pills ([28], p. 112). After all, the relation of beliefs and effective action is not logical, but empirical. In some instances consensus undoubtedly will lead to therapeutic effectiveness. It is sometimes thought that this is most likely the case in so-called mental or psychological illness. Yet when cure or amelioration of symptoms occur, it may be that the actual cause of the cure or remission is unrelated to social beliefs or that social beliefs are but one among many causally significant factors.

Consensus, instead of defining therapeutic power, functions more significantly in determining the *status* of the healer or practitioner.

> Community belief may provide the criterion for a social role – this is a logical or conceptual connection. On the other hand, community belief may produce the power associated with a social role – this may be logical (as with kings, all of whose powers are social), but more interestingly may be an empirical connection. That social role and power can be separated is shown, in the cases of doctors and patients, by the fact that people may be able to actually heal or actually suffer independently of the opinion of their society ([28], p. 112).

Thus, the distinction between modern and classical views of therapy may be misleading if it is taken either to indicate that medicine and the particularly modern physician-patient relationship are not supported by social consensus or that therapeutic power is a direct result of social consensus. Quite the contrary seems to be the case. It appears, however, that the consensus shaping medicine concerns precisely a recognition of the individual character and nature of therapy and a particular view of well-being and therapeutic power which is radically different from the classical view described by Rieff.

In medicine well-being is predicated on the general goal of rational free action. Hence, well-being is not defined positively, but privately in terms of the inability to exercise rational free action. This may be due partly to the absence of consensus regarding a particular positive definition of human well-being. But more importantly, it may be due to a recognition of the empirical and fallible character of all therapy. By all measures, negative definition of well-being or health in terms of disease categories appears to be primary in medicine. In terms of the goal of rational free action, those conditions or states which restrict the basis of such activity tend to count as diseases; those states which augment rational free action count as health ([9], p. 127).

Unlike other therapeutic activities medicine focuses on the preservation or re-establishment of human well-being understood in terms of empirical regularities, not in terms of moral blame and praise ([9], p. 121). The concrete definition of well-being is open to specification implicitly in terms of what humans generally desire at any historical period, but explicitly in terms of disease concepts. Definition of well-being in terms of disease thus logically commits medicine (and therapy in medicine) to explanation, prediction, and control. Expressed otherwise, it includes diagnosis, prognosis, and therapy. Here, 'therapy' (or 'therapeutic') means technical intervention or interaction predicated on *rational, scientific* understanding. This commitment distinguishes medicine from other therapeutic activities such as education and moral discourse ([9], p. 121). The functions of diagnosis, prognosis, and therapeutics thus comprise three main aspects of the physician-patient relationship. As Edmund D. Pellegrino and David C. Thomasma have argued: "Of the three forms . . . the last is the most important. . . . This form comes closest to the motives and curative intent of clinical interaction. What to do for the patient . . . is the epistemological guide for all aspects of the medical event" ([32], p. 71).

Judgments regarding what is to be done for a given patient interweave technical and moral dimensions.[7] On the classical view, moral dimensions – implicit normative community judgments regarding health – enjoy a primacy over technical, scientific considerations. What is done therapeutically is sanctioned by a community belief system without regard to the question of the efficacy or safety of procedures or interventions. To even raise concerns such as safety or efficacy requires the 'critical detachment' identified by Rieff. Medicine specifies the general definition of therapeutic relationships offered earlier in terms of beliefs in empirical, scientific knowledge and the value of rational free agency. Because disease language unifies these beliefs and values, it assumes – in diagnosis – a rather prominent place in physician-patient encounters.

Some of the implications of this concept of the therapeutic character of the physician-patient relationship can be examined by considering two rather different ways that the question has been addressed. Since both involve consideration of rules as well as beliefs and values, they will serve to amplify the earlier theoretical points regarding the distinction between rules and beliefs and values in the definition of a practice. The first involves an approach concerned with models of physician-patient encounters which confuses the structure of such relationships with normative beliefs and values; the second concerns the distinction between research and practice (or therapy).

MODELS OF THERAPY IN MEDICINE

Thomas Szasz and Marc Hollender have described three main types of physician-patient interactions: activity-passivity, guidance-cooperation, and mutual participation [41]. Because their classification is so well-known and discussed, it can provide a convenient locus for discussing the structure of therapeutic relations in medicine and its bearing on definitions of well-being and power. The *activity-passivity* model of the physician-patient relationship is one in which the physician does something to the patient and the patient is passive as in anesthesia, acute trauma, or coma; the *guidance-cooperation* model is a relationship in which the physician tells the patient what to do (doctor's orders) and the patient obeys as in acute infection; and the *mutual participation* model is a relationship in which the physician helps the patient to help himself and the patient uses expert help as in most chronic illnesses and psychoanalysis ([41], pp. 586–587). Szasz and Hollender approach these models on the basis of the psychology of the physician and criticize the activity-passivity and guidance-cooperation models.

These two models, they argue, primarily gratify the physician's needs for mastery and contribute to his feelings of superiority. In other words, these relationships are predicated on their psychological advantage to the physician. Szasz and Hollender instead favor the mutual participation model which they believe to be essentially foreign to medicine because it is characterized by a high degree of empathy and has elements often associated with friendship and partnership. On this model, the physician helps the patient to help himself. The physician's gratification does not stem from his power or from the control he exercises over his patient: "His satisfactions are derived from more abstract kinds of mastery, which are as yet poorly understood" ([41], p. 588).

Szasz and Hollender, however, recognize that the practice of medicine presents situations in which each of these models is appropriate and in which transitions from one to another seem to occur. The comatose patient, for example, will either recover consciousness or die. If he improves, the relationship with the physician must also change. But the change which they acknowledge is not in terms of the models so much as in physician *attitude:*

It is at this point that the physician's inner (usually unacknowledged) needs are most likely to interfere with what is "best" for the patient. At this juncture, the physician either changes his "attitude" (not a consciously or deliberately assumed role) to complement the patient's emergent needs or he foists upon the patient the same role of helpless passivity from which he (allegedly) tried to rescue him in the first place. Here we touch on a subject rich in psychological and sociological complexities. The process of change the physician must undergo to have a mutual constructive experience with the patient is similar to a very familiar process: namely the need for the parent to behave ever differently toward his growing child ([41], p. 588).

This approach leaves a number of important questions unanswered. For example, on what basis is the judgment regarding what is 'best' for the patient made? And why does the physician change only his 'attitude' rather than his role or the nature of the relationship in responding to the patient's emergent needs? To answer these questions, it is useful to consider the reasons for Szasz and Hollender's preference for the mutual participation model.[8]

Szasz and Hollender argue that in mutual participation the physician does not profess to know exactly what is best for the patient. Rather the search for the therapeutic good becomes the essence of the therapeutic interaction. In this relationship the patient's own experiences furnish indispensible information regarding what he defines as 'health' ([41], p. 589). The basis for preferring mutual participation therefore is offered as a definition of the therapeutic encounter itself. On Szasz and Hollender's view, the therapeutic encounter is properly understood in terms of the freedom of the individual patient to define the goal of therapy. Here, they have hit on something crucial to therapeutic relationships in medicine – namely their connection with freedom – but they misconstrue its proper sense.

Szasz and Hollender define well-being or health in therapeutic relationships in emotivist terms.[9] They argue that " 'good' and 'bad' are personal judgments, usually decided on the basis of whether or not the object under consideration satisfies us" ([41], p. 588). They do note that "in viewing the doctor-patient relationship we cannot conclude, however, that anything which satisfies – irrespective of other considerations – is 'good' " ([41], p. 588). According to them, what additionally contributes to the definition of good in the physician-patient relationship is the physician's technical skill, training, and available technology. While acknowledging these factors to be important, they nonetheless argue that 'good medicine' can be considered from many viewpoints. Their own "is limited to but one – sometimes quite unimportant – aspect of the contact between physician and patient" ([41], p. 588n). This modesty, however, represents more than a methodological restriction. Their criticism of the activity-passivity and guidance-cooperation models of the physician-patient relationship is part of an attempt (best exemplified in [39]) to construct a vision of therapeutic relationships on the basis of a radical individualism. Accordingly, the therapeutic relationship is viewed as a free-standing relationship between autonomous individuals which abstracts from all social connections.[10] As a consequence, they deal with the social practice supporting therapeutic encounters in medicine either in critical terms – as in Szasz's arguments that mental illness is a myth – or they simply ignore the way in which it provides an ambient setting for their preferred model of interaction [40].

For example, Szasz and Hollender regard the concepts of the normal, abnormal, symptom, and disease, as social conventions and on that basis infer their arbitrary character. As a result, the power and authority associated with disease concepts is assumed logically to restrict patients' freedom, since there is no rational foundation for the authority and power which disease language vests in the physician ([39]; [41], p. 591).

This view seems to be based on the assumption that only the isolated individual can decide what constitutes illness or what thwarts his freedom and that the power and knowledge of the physician – lodged in diagnosis and therapeutic techniques – is itself necessarily coercive. Szasz forgets that all sorts of discomforts, disabilities, and deformities can restrict freedom and that such restrictions are social in nature because human action is always embodied action in the ambient social world, the world of everyday life. On this alternative view, illness is seen as an ontological feature of embodied existence as such ([32], pp. 207–208). However by collapsing ontological considerations into considerations of rules articulated in terms of the structure of physician-patient relationships, Szasz and Hollender fail to see that the question of the scope of the therapeutic relationship cannot be answered adequately by addressing the level of rules alone. Szasz, in particular, confuses not only rules with values and beliefs, but elaborates a view of therapy which celebrates the individual in isolation from social relationships. The defining limits of therapeutic relationships thus are determined *only* internally, in particular relationships.

RESEARCH AND THERAPY

Since research has been regarded as posing ethical problems requiring specific review and judgment, it is important to distinguish research and practice in order to determine which activities are subject to review requirements. The practice of medicine, however, has come to incorporate research activities –

in terms of what Talcott Parsons has termed "the professional complex" of research, service, and teaching – to such an extent that this distinction has proven troublesome [31]. This prompts the conceptual question regarding the connection of research and therapy and the implications of research for therapeutic interactions [6]. Fortunately, this question can be approached by considering two reasons which make distinguishing research and therapy difficult: the confusion surrounding the term 'therapeutic research' and the 'experimental' nature of medical practice.

It is not clear how the term *therapeutic research* entered usage, but it has certainly caused difficulty. Since it is not done for the benefit of the subject, non-therapeutic research is often seen as suspect ethically and as requiring stronger justification than therapeutic research which, it is often implied, is of therapeutic benefit to the subject.[11] This implication, however, is not valid because it assumes what is precisely in question. The objective of research usually is to determine the efficacy of a particular therapeutic method or procedure. Implicit in the need to ask the research question is a basic scientific uncertainty regarding which method or procedure provides the best outcome. This uncertainty means that the question of therapeutic benefit is at issue in the research and so cannot be validly assumed ([35], p. 2).

Robert J. Levine and Karen Lebacqz, both members of the National Commission for the Protection of Human Subjects of Biomedical and Behavioral Research, have separately criticised the National Commission's early definition of 'therapeutic research' or "the spurious distinction between therapeutic and non-therapeutic experimentation" in its first report, *Research on the Fetus* ([20], [22], [27]). Both argue for a view later articulated in the National Commission's *The Belmont Report:*

> For the most part, the term "practice" refers to interventions that are designed solely to enhance the well-being of an individual patient or client and that have a reasonable expectation of success. The purpose of medical or behavioral practice is to provide diagnosis, preventive treatment or therapy to particular individuals. By contrast, the term "research" designates an activity designed to test a hypothesis, permit conclusions to be drawn and thereby to develop or contribute to generalizable knowledge (expressed, for example, in theories, principles, and statements of relationships) ([26], pp. 2–3).

On this view, research and practice (or 'therapy') are logically distinct. Research is a class of activities designed to produce generalizable knowledge. As such it is fundamentally different from treatment or therapy. The distinction is predicated primarily on the intent of both activities which is reflected in their planning, execution, and the structure of their relationships. While research may be conjoined with therapy, as in research on the treatment of individuals, the research itself does not thereby become 'therapeutic' ([20], p. 362). Confusion occurs because research and therapy can occur together in the practice of medicine – as in research designed to evaluate the safety and efficacy of a therapeutic method or procedure – and because notable departures from standard practice are often called 'experimental' when the terms 'experimentation' and 'research' are not carefully defined ([26], p. 2).

Levine argues that departures from standard practice, which *The Belmont Report* terms 'innovations', should be thought of as a class of activities separate from research and those therapies which have been scientifically validated and established. He terms this class *non-validated practices*: "novelty is not the attribute that defines this class of practices; rather it is the lack of suitable validation of the safety or efficacy of the practice" ([21], p. 22). In *The Belmont Report*, the National Commission focused on innovations:

> When a clinician departs in a significant way from a standard or accepted practice, the innovation does not, in and of itself, constitute research. The fact that a procedure is "experimental" in the sense of new, untested or different, does not automatically place it in the category of research. Radically new procedures of this description should, however, be made the object of formal research at an early stage in order to determine whether they are safe and effective ([26], p. 3).

Requiring that innovations or non-validated practices be made the object of research raises the converse and more important question for present purposes: what is the proper function of 'experimentation' in the practice of medicine? Further, can experimentation and research be included within the legitimate scope of the therapeutic relationship in medicine?

To answer these questions, it is necessary to be clear about the meaning of 'experimentation'. An experiment is not only a test of something, but also is a tentative procedure adopted under conditions of uncertainty as to whether it will achieve the desired purposes or results ([21], p. 21n). On this definition, much of the practice of medicine is experimental in nature ([5], p. 44; [34], p. 358). This view of the experimental nature of therapy depends on a recognition that medicine as a practice functions and is structured on the basis of the diversity of individuals and that there are inherent limitations to generalization and predictive power. Medicine, thus, is founded on uncertainty, an uncertainty that permeates all aspects of the therapeutic relationship. Samuel Gorovitz and Alasdair MacIntyre have argued that an important implication of this point concerns the fallibility of medical practice:

All experiments necessarily involve the possibility of failure in the sense that the expected or hypothesized outcome may not occur; whereas other outcomes, unintended and not usually specifiable entirely in advance, may occur. Thus the possibility of failure, and even of damaging failure, is linked conceptually – and not merely contingently – to the notion of experimentation, and therefore to the practice of clinical medicine ([14], p. 265).

Knowledge of the individual is not only essential to medical practice, but is always potentially inadequate to the extent that a damaging error may result from otherwise conscientious and well-motivated actions ([14], p. 266). Experimentation thus turns out to be an essential feature of therapy owing to the necessary fallibility of therapeutic judgment in light of the purpose of the therapeutic relationship to improve or contribute to the well-being of the patient.

The relationship of research and therapy is therefore intimate. Because medicine is an empirical, scientific practice, therapy necessarily implicates research in the effort to discover safe and effective treatments. Experimentation and therapy are similarly linked conceptually insofar as medicine aims at individual well-being under conditions of uncertainty. The aim or goal of therapy in medicine thus has two aspects: "It is the choice of what is *right* in the sense of what conforms scientifically, logically, and technically to the patient's needs, and a choice of what is *good*, what is worthwhile for the patient" ([32], p. 311). As a practice predicated on the value of rational free action, medicine necessarily incorporates research and experimentation as features of a scientifically-based therapy.

IMPLICATIONS

The concept of the therapeutic relationship in medicine must be interpreted in terms of the practice of medicine. Within that practice therapeutic relationships are constituted at two logically distinct levels: the level of rules and the level of values and beliefs. The present essay has focused on the theoretical or normative function of beliefs and values which most generally define therapeutic relationships in medicine. This focus should not be taken to imply that values and beliefs alone adequately define the scope of therapy in medicine. Quite to the contrary, the more comfortable level of rules and its affiliated concepts of authority, power, and rights must be incorporated in any full philosophical account. Nevertheless, given the variety of therapeutic relationships, it is important first to be clear about medicine's most general distinguishing features.

Focusing on the values and beliefs which structure the practice of medicine aids in demarcating the conceptual limits of therapeutic relationships, but of course does not clarify the actual or ideal nature of therapeutic encounters in medicine. Such clarification would involve incorporation not only of beliefs and values with rules, but justification of beliefs, values, and rules in actual or ideal relationships. Although the present essay has eschewed justification, it does have implications for it. Because justification properly involves beliefs and values, it cannot be restricted to the level of rules. Without an analysis of the basic values and beliefs underlying the practice of medicine, normative arguments for particular models of therapy can be confused with descriptive assessments of actual or possible relationships. Justification properly understood would incorporate the rules which determine the content and structure of therapeutic relationships with medicine's overall commitment to a rational, scientific understanding of treatment and the goal of rational, free action. Without explicit reference to these broader commitments, arguments for particular models of therapy cannot be adequately justified.

This view of justification of the scope of therapy in medicine entails that analysis of problems or conflicts at the level of rules can be made complete only by explicit reference to the concept of the practice of medicine. General ethical and epistemological principles of medicine thus become central to particular ethical dilemmas, issues, or problems arising at the level of rules. Questions regarding the scope of therapy and therapeutic relationships in medicine therefore implicate philosophical questions about the content and structure of medicine as a social practice and open the prospects for a philosophical critique of medicine.

Southern Illinois University School of Medicine,
Springfield, Illinois

NOTES

[1] Similarly, ethical questions involving concepts of patient autonomy and freedom already presume a certain conceptual model or theory of therapy and therapeutic relationships. On the other hand, some of the criticism of the patient rights movement and medical ethics have come from practitioners who urge that the exclusive concern for patient rights in abstraction from the requirements of treatment can actually thwart therapeutic relationships and, as a consequence, result in absurd visions of patient rights ([4], [15]). Admittedly, there are political elements and motivations implicit in these arguments. But it also must be noted that philosophical assumptions regarding therapeutic relationships have certainly been inadequately addressed in their own terms. It is the

task of this essay, then, to provide a framework for addressing some of these issues by sketching a theory of therapeutic relationships in medicine.

[2] My discussion of practices and medicine as a practice relies heavily on the work of Richard E. Flathman ([10], [11], [12]).

[3] This admixture of administrative, bureaucratic, and therapeutic functions comprises one of the more serious problems for medical ethics. Some of the issues involved have been addressed under the rubric of responsibility ([3], [29]).

[4] Critics of medicine and health care often focus on individuals and often propose *educational* reforms. As Eliot Freidson critically notes: "In considering how the members of the profession work its leaders typically see solutions to the problem of poor or unethical work in recruiting better-motivated and more capable entrants to school, in improving their professional education and in generally 'raising standards'. All these devices are predicated on the aim of changing the quality of individuals, the assumption being first that social pathologies connected with medical care, like illnesses connected with mankind, are 'caused' by the characteristics of the individual providing the care rather than by the environment in which those individuals provide care, and second that they are best treated by treating the individual rather than the environment" ([13], p. 6).

[5] Alasdair MacIntyre offers the most extended critique of the philosophical consequences of the severing of community and morality in *After Virtue* [23]. One result is the way that the concept of the therapeutic has been used to displace truth as a value in the interest of psychological effectiveness ([23], p. 29). See also Sigmund Koch's 'The Image of Man in Encounter Group Theory' [18].

[6] The view that concepts of health and disease are value-laden and socially determined does not entail that they are capricious or arbitrary. The meaning of disease language is culture-specific, but illness is an ontological feature of finitude. There is a transcultural basis for disease [25]. Disease language relates to illness as the attempt to explain, predict, and control certain states of illness. Viewed generally, then, disease language is one way of identifying and dealing with illnesses which have special social significance.

Since therapeutic relationships are social relationships involving special social roles and role-defined responsibilities and actions, a key question concerns how social values and beliefs influence or contribute to the existence, function, and meaning of authority, power, and rights in the practice. In part, this question concerns the nature of authority and power and their relation to social consensus. For a theoretical discussion of this question, see [12], [16], [42].

[7] Throughout the history of medicine, but especially in the modern period, the terms *therapy* and *therapeutics* have had a technical flavor and focus. See, for example, Erwin Ackerknecht [1]. But technical power alone cannot suffice. If one regards therapeutic techniques on a spectrum from those that involve freedom and thought to those that do not, one can question whether those techniques or treatments which are at the nonthought end are truly 'therapy'. Jerome Neu has argued in this regard: 'Torture may produce psychological and behavioral changes, but it is not to be regarded as 'therapy'. This is partly because of the content of the treatment (pain, etc.), but more importantly because of its purposes and the desires of the person who initiates it (who is *not* himself the patient or victim). It may also be in part because of the mechanisms involved. And if, for example, terror of the treatment itself is part of what contributes to nonthought 'cures', we may not wish to regard the changes as 'therapy'. Perhaps the terror of treatment forces the patient to overcome the overt manifestations of his

problem. But is that 'cure' (even if the manifestations disappear permanently)?" ([28], p. 104).

On the side of freedom and thought, the question of therapeutic technique points to psychoanalysis. Freud and his followers saw their problem as reestablishing the effectiveness of verbal (generally oral) therapy in the milieu over-reliant on the purely somatic. This contrasts with the ancient Greeks for whom the problem was to establish what might be truly therapeutic in the welter of verbal activities generated by a culture where both mores and academic education were rhetorical ([30], p. x).

[8] Preferring a relationship similar to the parent-child relationship is a tacit recognition of the problem of responsibility in medicine, a problem partly due to the highly complex institutional character of modern medical practice. In such settings, relationships between physicians and patients can take on the anonymity characteristic of encounters with strangers. And so any relationship which cultivates a personal interaction seems intuitively preferable. For this reason the idea of the parent-child relationship offers more than solace. Hans Jonas has argued that the parent-child relationship is the paradigm for the concept of responsibility [12]. In the case of medicine, John Ladd has gone so far as to argue that responsibility relationships must be *personal* rather than role-defined [19]. I have criticized this latter approach as violating the social and institutional sense of medicine in an essay entitled 'The Concept of Responsibility in Medicine' [2].

[9] MacIntyre's criticism of emotivism in *After Virtue* is especially compelling ([23], pp. 6–34).

[10] Szasz and Hollender treat the physician-patient relationship not as a social role relationship, but as a 'human relationship' between individuals. But a human relationship on their view has no reality; it is "an abstraction appropriate for the description and handling of certain observational facts" ([41], p. 585).

[11] In fact, a proposed, but never implemented policy of DHEW involved prohibiting entire categories of 'non-beneficial research' without regard to consideration of risk involved [8].

BIBLIOGRAPHY

1. Ackerknecht, E. H.: 1973, *Therapeutics: From the Primitives to the 20th Century*, Hafner Press, New York.
2. Agich, G. J.: 1982, 'The Concept of Responsibility in Medicine', in G. J. Agich (ed.), *Responsibility in Health Care*, D. Reidel Publishing Company, Dordrecht, pp. 53–73.
3. Agich, G. J. (ed.): 1982, *Responsibility in Health Care*, D. Reidel Publishing Company, Dordrecht.
4. Appelbaum, P. S. and Gutheil, T. G.: 1979, 'Rotting with Their Rights On: Constitution Theory and Reality in Drug Refusal by Psychiatric Patients', *Bulletin of the American Journal of Psychiatry and the Law* 7, 308–317.
5. Blumgart, H. L.: 1970, 'The Medical Framework for Viewing the Problem of Human Experimentation', in P. A. Freund (ed.), *Experimentation with Human Subjects*, George Braziller, New York, pp. 39–65.
6. Churchill, L. R.: 1980, 'Physician-Investigator/Patient-Subject: Exploring the Logic and Tension', *The Journal of Medicine and Philosophy* 5, 215–224.

7. Cushman, R. E.: 1958, *Therapeia: Plato's Conception of Philosophy*, The University of North Carolina Press, Chapel Hill.
8. Department of Health, Education, and Welfare: 1974, 'Proposed Policy', *Federal Register* 39, 30648–30657.
9. Engelhardt, Jr., H. T.: 1976, 'Human Well-Being and Medicine: Some Basic Value Judgments in the Biomedical Sciences', in H. T. Engelhardt, Jr. and D. Callahan (eds.), *Science, Ethics and Medicine*, Institute of Society, Ethics and the Life Sciences, Hastings-on-Hudson, New York, pp. 120–139.
10. Flathman, R. E.: 1980, *The Practice of Political Authority: Authority and the Authoritative*, University of Chicago Press, Chicago.
11. Flathman, R. E.: 1976, *The Practice of Rights*, Cambridge University Press, New York.
12. Flathman, R. E.: 1982, 'Power, Authority, and Rights in the Practice of Medicine', in G. J. Agich (ed.), *Responsibility in Health Care*, D. Reidel Publishing Company, Dordrecht, pp. 105–125.
13. Freidson, E.: 1970, *Professional Dominance: The Social Structure of Medical Care*, Atherton Press, New York.
14. Gorovitz, S. and MacIntyre, A.: 1976, 'Toward a Theory of Medical Fallibility', in H. T. Engelhardt, Jr. and D. Callahan (eds.), *Science, Ethics and Medicine*, Institute of Society, Ethics and Life Sciences, Hastings-on-Hudson, New York, pp. 248–284.
15. Gutheil, T. G. and Appelbaum, P. S.: 1980, 'The Patient Always Pays: Reflections on the Boston State Case and the Right to Rot', *Man and Medicine* 5, 3–11.
16. Hauerwas, S.: 1982, 'Authority and the Profession of Medicine', in G. J. Agich (ed.), *Responsibility in Health Care*, D. Reidel Publishing Company, Dordrecht, pp. 83–104.
17. Jonas, H.: 1977, 'The Concept of Responsibility: An Inquiry into the Foundations of an Ethics of Our Age', in H. T. Engelhardt, Jr. and D. Callahan (eds.), *Knowledge, Value and Belief*, Institute of Society, Ethics and Life Sciences, Hastings-on-Hudson, New York, pp. 169–198.
18. Koch, S.: 1972, 'The Image of Man Implicit in Encounter Group Therapy', in J. D. Matarazzo *et al.* (eds.), *Psychotherapy 1971*, Aldine-Atherton, Inc. New York and Chicago, pp. 535–555.
19. Ladd, J.: 1978, 'Legalism and Medical Ethics', in J. W. Davis *et al.* (eds.), *Contemporary Issues in Biomedical Ethics*, Humana Press, Clifton, New Jersey.
20. Lebacqz, K.: 1977, 'Reflections on the Report and Recommendations of the National Commission: Research on the Fetus', *Villanova Law Review* 22, 357–366.
21. Levine, R. J.: 1979, 'Clarifying the Concepts of Research Ethics', *The Hastings Center Report* 9 (June), 21–26.
22. Levine, R. J.: 1977, 'The Impact on Fetal Research of the Report of the National Commission for the Protection of Human Subjects of Biomedical and Behavioral Research', *Villanova Law Review* 22, 367–383.
23. MacIntyre, A.: 1981, *After Virtue*, University of Notre Dame Press, Notre Dame.
24. Moore, F. D.: 1970, 'Therapeutic Innovation: Ethical Boundaries in the Initial Clinical Trial of New Drugs and Surgical Procedures', in P. A. Freund (ed.), *Experimentation with Human Subjects*, George Braziller, New York, pp. 358–378.
25. Murphy, J. M.: 1976, 'Psychiatric Labeling in Cross-Cultural Perspective', *Science* **191**, 1019–1028.

26. National Commission for the Protection of Human Subjects of Biomedical and Behavioral Research: 1978, *The Belmont Report: Ethical Principles and Guidelines for the Protection of Human Subjects of Research*, DHEW Publication No. (OS) 78–0012, Washington.
27. National Commission for the Protection of Human Subjects of Biomedical and Behavioral Research: 1975, *Report and Recommendations: Research on the Fetus*, DHEW Publication No. (OS) 76–127, Washington.
28. Neu, J.: 1976, 'Thoughts, Theory, and Therapy', in D. Spence (ed.), *Psychoanalysis and Contemporary Science*, International Universities Press, Inc., New York, pp. 103–143.
29. Newton, L. H. and Pellegrino E. D. (eds.): 1982, 'Collective Responsibility in Medicine', *The Journal of Medicine and Philosophy* 7.
30. Ong, W. J.: 1970, 'Forward', in Pedro Laín Entralgo, *The Therapy of the Word in Classical Antiquity*, L. J. Rather and J. M. Sharp (eds. and trs.), Yale University Press, New Haven, pp. ix–xvi.
31. Parsons, T.: 1970, 'Research with Human Subjects and the "Professional Complex" ', in P. A. Freund (ed.), *Experimentation with Human Subjects*, George Braziller, New York, pp. 116–151.
32. Pellegrino, E. D. and Thomasma, D. C.: 1981, *A Philosophical Basis of Medical Practice*, Oxford University Press, New York and Oxford.
33. Rieff, P.: 1966, *The Triumph of the Therapeutic: Uses of Faith after Freud*, Harper and Row Publishers, New York.
34. Risse, G. B.: 1982, 'Once on Top, Now on Tap: American Physicians View Their Relationships with Patients, 1920–1970', in G. J. Agich (ed.), *Responsibility in Health Care*, D. Reidel Publishing Company, Dordrecht, pp. 23–49.
35. Rolleston, F. and Miller, J. R.: 1981, 'Therapy or Research: A Need for Precision', *IRB: A Review of Human Subjects Research* 3 (August–September), 1–3.
36. Siegler, M.: 1982, 'Medical Consultation in the Context of the Physician-Patient Relationship', in G. J. Agich (ed.), *Responsibility in Health Care*, D. Reidel Publishing Company, Dordrecht, pp. 141–162.
37. Simon, B.: 1978, *Mind and Madness in Ancient Greece: The Classical Roots of Modern Psychiatry*, Cornell University Press, Ithaca and London.
38. Strong, P. M.: 1979, *The Ceremonial Order of the Clinic*, Routledge and Kegan Paul, London.
39. Szasz, T. S.: 1974, *The Ethics of Psychoanalysis*, Basic Books, Inc. Publishers, New York.
40. Szasz, T. S.: 1961, *The Myth of Mental Illness: Foundation of a Theory of Personal Conduct*, Hoeber-Harper, New York.
41. Szasz, T. S. and Hollender, M. H.: 1956, 'A Contribution to the Philosophy of Medicine: The Basic Models of the Doctor-Patient Relationship', *AMA Archives of Internal Medicine* 97, 585–592.
42. Veatch, R. M.: 1982, 'Medical Authority and Professional Authority: The Nature of Authority in Medicine for Decisions by Lay Persons and Professionals', in G. J. Agich (ed.), *Responsibility in Health Care*, D. Reidel Publishing Company, Dordrecht, pp. 127–137.

SECTION IV

MORALITY IN THE PATIENT-PHYSICIAN RELATIONSHIP

H. TRISTRAM ENGELHARDT, JR.

THE PHYSICIAN-PATIENT RELATIONSHIP IN A SECULAR, PLURALIST SOCIETY

Medicine is practiced in secular, pluralist societies. In the United States and large areas of Latin America, Europe, and Japan, physicians and nurses find that their patients and colleagues are often moral strangers to them. They frequently do not share with their patients a common view of the good life or of the purposes and goals of health care. It is this circumstance that marks the moral predicament of the modern physician-patient relationship. This is not to deny that in many instances physicians do share with their patients common moral viewpoints or understandings. A Hassidic Jewish physician practicing in a Hassidic Jewish community, a Seventh Day Adventist in a Seventh Day Adventist hospital, a Roman Catholic practicing in a strict Roman Catholic hospital may be able to presume that his or her views concerning the fundamental nature of morals are shared with patients. That presumption, however, can usually not be made. Physicians and patients are likely to hold diverging views with regard not only to the moral probity of such widely employed interventions as contraception and abortion, but also with regard to the moral significance of pain, suffering, death, and life itself. Since concrete moral judgments require an appeal to a moral hierarchy of benefits and banes, differences in such hierarchies frame differences in world viewpoints and the significance of particular decisions.

This circumstance has only in part been recognized, and usually with reluctance. The ideal physician in 19th century America was a Christian physician ([5], pp. 21–25), and even into the current decade the prevailing moral orthodoxy is that of the Judeo-Christian tradition, though increasing numbers of the house staffs of American hospitals are Moslems, Hindus, Buddhists, and atheists.[1] One can in part attribute the growth of modern bioethics to a recognition of this cultural problem: bioethical issues arise in a context no longer informed by a single concrete moral viewpoint. The problem has been to develop a neutral moral language, which would allow moral discourse spanning communities of Catholics, Protestants, Jews, Moslems, Hindus, Buddhists, atheists, and humanists of particular persuasions. This moral lingua franca has been sought as a logic for a pluralism. It has been pursued as a neutral moral cement that would allow physicians and patients who did not share common moral ideals to work together within a

Earl E. Shelp (ed.), The Clinical Encounter, 253–266.

peaceable moral framework. It has been sought as a basis to allow them to meet at least as peaceable strangers.

In giving this sketch of bioethics, one must note that it for the most part captures only one dimension of the endeavor. Philosophical bioethics must be contrasted with the various bioethics of the particular moral communities which have endeavored to frame their own moral language for coming to terms with past moral problems in new settings, or with the new problems of modern technology. One might think here of Roman Catholic medical ethics, Jewish medical ethics, or Marxist medical ethics. These two elements of bioethics underscore the ambiguity of the physician-patient relationship. Physicians and patients meet as (1) moral strangers and also (2) moral friends. By the first I will mean that sort of relationship that physicians and patients have with each other when they confront a moral problem in medicine (e.g., when to perform prenatal diagnosis and then possibly an abortion, or when to cease 'heroic' care for a deformed neonate, and what to characterize as 'heroic'), without sharing a common concrete moral viewpoint. By the second I will mean that sort of relationship that physicians and patients have with each other when they confront a moral problem in medicine and share a concrete moral viewpoint (e.g., are both believing Roman Catholics, or committed humanists of some particular persuasion).

Though there is much to be said with regard to the relationship of physicians and patients as friends, as individuals sharing a common view of the good life, I will in this paper address primarily the issue of health care among strangers. In doing this, I will attempt to sketch not only the problems involved in health care in secular pluralist societies such as ours. I will additionally attempt to indicate the character of the solution to moral quandaries offered within such societies. Further, I will suggest that the character of secular pluralist societies, in particular with regard to health care, exists as a form of intellectual solution to the problem of moral discourse among individuals drawn from divergent communities of moral commitment.

THE SECULAR, PLURALIST, PEACEABLE SOCIETY

Most modern societies are secular in the sense that their major official public policy declarations are put in a language that does not appeal to a particular religion or religious understanding. Modern democracies are marked by a wide range of important moral communities including not only divergent varieties of Christians, Jews, Moslems, and Hindus, but Marxists, atheists, and others. Nation-states compass numerous moral communities with their divergent moral viewpoints. Further, these societies, even when they have not been able to reach agreement with regard to the moral probity of particular actions in health care, have espoused a more fundamental commitment to toleration and the peaceable negotiation of moral intuitions. The application of rules of free and informed consent to choices regarding the use of contraception and abortion are instructive examples. Where a concrete moral policy cannot be developed with authority, physicians and patients can meet to negotiate with each other their conjoint actions. Which is to say, moral discourse between physicians and patients who meet as strangers will not occur as it has often in the past, within the terms of a particular orthodoxy imposed upon the participants. (Consider, for example, a Jewish physician meeting with a Christian patient in Christian Europe of the Middle Ages. The language of moral discourse in resolving issues of bioethical conflict was presumed to be that of the orthodox Christian church.[2]) It now, instead, occurs in a language that is secular, and which presumes the presence of numerous and competing moral viewpoints, and which in addition presumes the moral importance of resolving issues peaceably (i.e., without the imposition of unconsented-to force).[3]

An account of the moral lineaments of the physician-patient relationship in a secular society will then have to include the notion of a peaceable community in order to provide an understanding of the moral fabric binding strangers in its physician-patient relationships. Indeed, a great proportion of modern bioethics has concerned itself with understanding the procedure for mutual negotiation among physicians and patients in such circumstances. A paradigm example is the development of canons for free and informed consent that do not impose an understanding of what well-informed individuals should hold to be the *best* choices. Rather, one finds the accent falling upon providing that information that free individuals are likely to wish to consider in their free choices. The accent moves from content to fair procedure, from an accent upon professional standards regarding the amount and character of information to be given to patients, to standards dependent upon insuring that free individuals can choose freely, though not necessarily well ([17], [18], [19], [20]).

In addition to ensuring fair procedures, physicians function as geographers of value. Even where physicians may not impose a particular view of the good life upon their patients, they can at least enable patients to choose well in their own terms. Thus, in addition to respecting freedom as a side-constraint upon physician-patient relations (as the minimal condition for preserving the peaceable society), there is often an additional focus of attention on the good of the patient as the patient conceives it.[4] A reconstruction of the moral commitment that binds physicians and patients together in a secular pluralist society will, then, disclose not only respect for autonomy, but often a very general principle of beneficence, of doing to others their good. One will note, however, how this lack of content changes the usual character, for example, of the Golden Rule. As a principle respecting autonomy, the rule would need to read something like this: "Do not do unto others what they would not have done unto themselves, and do with them that which is mutually agreed." The principle of beneficence recast to mimic the Golden Rule would read: "Do to others their good." The first principle, since it turns on the very notion of a peaceable community, is more binding. Where physicians and patients meet as strangers in a secular pluralist society, each possesses a moral trump, for either can refuse to enter into a relationship when there is moral disagreement. In respecting this principle, the minimum condition of a peaceable society, in the sense of a society not based on the use of force against the unconsenting innocent, is fulfilled. The principle of beneficence would appear to be more supererogatory. Thus, for example, in terms of a secular pluralist society, if the only physician in a small town is a devout Roman Catholic, and if he or she provides abortions to patients when in need, such actions are likely to be seen as actions of beneficence, though surely not obligatory.

BIOETHICAL CONTROVERSIES

The moral and intellectual context of a secular pluralist society sets special problems for the resolution of moral controversy in the physician-patient relationship. In general, disputes about the moral probity of particular lines of conduct can be resolved in four ways: (1) they can be resolved by appeal to force. Usually the physician is in an advantaged position and can bring cloture to a debate. However, such an appeal to force does not resolve the ethical issue as an intellectual one. One turns to philosophical reflection in order not to determine who is likely to win a contest, but to find grounds for holding that a particular moral viewpoint is justified. (2) An appeal can be made to a set of commonly held moral beliefs, even if these are not amenable to general rational justification. Thus, for example, Jehovah's Witness physicians and patients can resolve for themselves moral disputes with regard to the use of blood products by an appeal to articles of faith and the religion's established means for mediating religious controversies, even if rational individuals generally may not be persuaded of the rational bases of the religious claims. When physicians and patients meet as strangers, they meet under circumstances where such appeals will not succeed. (3) An appeal can also be made to rational arguments. However, ethics appears to be at a special disadvantage. Disputes in mathematics or logic can be resolved by appeals to interests in coherence and consistency. Controversies in the empirical sciences can at least partly be resolved by appeal to the 'facts', even though the facts come always to a greater or lesser degree clothed in the historically, socially, and psychologically conditioned expectations of scientists.[5] To resolve a moral claim, however, one must appeal to a particular moral sense. This can be done by an appeal to particular intuitions, to the viewpoint of a disinterested observer, to a set of hypothetical contractors, or the discerning moral sense of an individual delineating natural law or the moral content of reason. However, for the intuiter, ideal observer, hypothetical contractor, or exponent of natural law to choose correctly, he or she must ground that choice in a particular moral sense. The difficulty is to know how to choose the correct moral sense. An appeal to a further, higher order moral sense would launch one on an argument of infinite regression. Rational arguments thus appear impotent to resolve controversies that turn on concrete understandings of the good life or the proper goals of health care. There is, however, (4) the possibility of negotiating a peaceable resolution to a controversy. As such, the core of ethics becomes that of mutual respect of the moral agents involved in a controversy. Again, this is why free and informed consent becomes the paradigm ethical issue in a secular pluralist society. Since the Golden Rule, "Do unto others as you would have them do unto you" becomes an invitation to moral tyranny, one accepts instead the principle of "Do not do unto others what they would not have done unto themselves, and do with them that which is mutually agreed." The concrete fabric of the moral life and of the physician-patient relationship in a secular pluralist society is therefore to be invented by appeal to a procedure of mutual respect of the freedom of the participants in the dispute.

An examination of the possibility of resolving controversies between physicians and patients thus leads one to a cardinal element of the relationship between strangers in a secular pluralist society. Their concrete moral character is fashioned by formal or informal negotiation. This is the case if for no other reason than that no other justifiable avenue is available. An appeal to force is not only intellectually unjustifiable, it is contrary to the very possibility of a peaceable society. An appeal to a common framework of belief is factually impossible, given the regnant pluralism of such societies.

Further, reason appears to be unable to deliver sufficient arguments definitively to resolve moral controversies except to reject those viewpoints which support the use of unconsented-to force against innocents. For example, though it is clear that experimentation on unconsenting, competent adults is incompatible with the very notion of a peaceable society, it is not at all clear whether abortion, the use of contraception, or allowing deformed neonates (who are not yet moral agents in *sensu stricto*) to die, should count as immoral actions. The impropriety of such actions depends upon less clearly justifiable arguments regarding the status of fetuses and infants, as well as the moral goods at stake in health care. However, even if a concrete moral understanding of the goods at stake in the physician-patient relationship can not be discovered, the notion of a peaceable society delivers at least a general formal constraint in terms of which one can create content for such a relationship. As a minimal notion of ethics, it is the one which is dependent least upon particular views of the good life. It requires only a commitment to resolving disputes not on the basis of force, but on the basis of reasons, considerations, and peaceable manipulations.[6] It provides the grammar of moral discourse through a procedure of mutual respect. When moral strangers meet, it allows them to act in mutual respect, if not agreement. In so doing, it provides them with a means for peaceably creating a common frame of action. Physicians and patients together, when they meet as strangers, must fashion the concrete moral structure of their relationship.

THE PHYSICIAN AS BUREAUCRAT AND GEOGRAPHER

In a recent volume, Alasdair MacIntyre speaks critically of the current bureaucratic individualism of the West. He sees it as a reaction to a flawed condition in which there is no longer a moral consensus, and therefore no longer the possibility of understanding with clarity the virtues of the roles one plays in society. "In any society where government does not express or represent the moral community of the citizens, but is instead a set of institutional arrangements for imposing a bureaucratised unity on a society which lacks genuine moral consensus, the nature of political obligation becomes systematically unclear" ([15], p. 236). This characterization by Alasdair MacIntyre is accurate. It is descriptive of a cardinal mark of a peaceable, pluralist society. If a society spans numerous moral communities, the moral cement that will hold it together will take on the character of general bureaucratic regulations. The fabric will reflect procedural guarantees of the freedom of those involved, including specially created entitlements and constraints.

This moral fabric, however, will not express the commitments of a particular moral viewpoint or of a particular moral community. If it did, then the fabric of a pluralist society would have come to be a vehicle for the imposition of the moral viewpoint of one of the particular moral communities constituting the pluralism upon the rest.

One might think here of the rules and regulations fashioned to safeguard free and informed consent. On the one hand, they reflect a general abstract understanding of what it is for free individuals to treat each other with mutual respect. On the other hand, as in the case of the rules governing research with federal funds, one finds special constraints created by the individuals who constitute the society [6]. They do not necessarily express a general moral sense or consensus, but a political compromise fashioned in order to allow individuals to reach across moral communities. The regulations concerning fetal experimentation ([6], Sections 46.201–46.211) might be helpful as an example in this regard, in that large proportions of American society would disapprove of such research, while others might see no moral basis for most of the regulations imposed [7]. The result has not been a proscription of such research itself, at least on a federal level, but only the erection of constraints regarding the use of commonly produced federal funds.

In secular pluralist societies physicians assume the role of bureaucrats in the sense of being custodians of procedures of fair interaction among free individuals, including special constraints that free individuals have fashioned. The circumstance tends to be bureaucratized in that formal and explicit statements of procedures for fairness are required the more the participants are strangers to each other. As strangers, less can be presumed implicitly, and more must be stated explicitly. A bureaucracy of formal statements for fair procedure in the form of explicit rules, along with custodians and interpreters of those rules, becomes inevitable. When one is not able to live in a community of friends where virtues and roles can be known in their concreteness without explicit statement, one is forced to embrace the bureaucratization of society. A degree of bureaucratization is the unavoidable cost of protecting strangers when they meet for health care.

As Alasdair MacIntyre indicates, this leads as well to an individualism. But this, too, is unavoidable. In a secular pluralist society, authority can be derived only from the concurrence of individuals, from the consent of those who participate. Modern democracies tend towards this account of authority, not necessarily out of choice, but in the absence of (1) the capacity of reason to justify conclusively one moral viewpoint, (2) all persons embracing one moral viewpoint, or (3) a coercive force compelling all to conform to one moral viewpoint. These three absences which characterize the condition of secular, pluralist, peaceable societies, characterize as well the limits on authority for the physician in physician-patient relationships in such societies. In the absence of a single, concrete moral viewpoint being available to condition physician-patient relationships, physicians and patients will need to fashion together a joint understanding of their undertaking. In this endeavor the physician is at best an authority, though without authority to impose particular choices. Insofar as one can speak of the physician's being in authority, such a role is shared with the patient in protecting procedures for fair negotiation [12].

Further, in order to work effectively with patients and for their good, physicians will need to disclose to patients the likely costs of their decisions in terms of the moral viewpoints of those patients. Since physicians see more of the usual outcome of decisions with regard to treatment and medical intervention than most patients do, they tend to become expert in assessing (i.e., authorities concerning) the likelihood of particular events. However, the significance of those events can be assayed only in terms of a particular moral viewpoint. For a physician to work with a particular patient, the physician must then chart with the patient the significance of the effects of treatment and intervention within the patient's moral framework. Physicians are thus forced to learn the geography of moral viewpoints in which they do not live. Constrained by their neutrality as bureaucrats, they can function at most as travel guides, suggesting that perhaps certain choices are better from the physician's point of view than others. Again, physicians are authorities, but not *in* authority to impose a particular choice. Their authority to act comes from the consent of their patients.

If I may summarize to this point, the fact that health care is delivered in societies spanning numerous moral communities, leads to physicians' playing a double role in physician-patient relationships. First, the physician must meet patients fairly, though the patients may be strangers. This will require attending to procedures of fair negotiation. Since strangers may not share a moral viewpoint, these procedures must be spelled out explicitly, and steps must be taken to assure mutual understanding. Formal rules for behavior appear inevitable. This element of the physician's role I have termed the role of the physician as bureaucrat. The physician must also aid actual patients in particular decisions framed in terms of the patient's particular moral viewpoint. Insofar as such viewpoints do not contradict the lineaments of a peaceable society, the physician can display the costs of alternative choices and suggest the virtue of particular choices. The physician, however, may not compel a particular choice. This role I have characterized as that of the physician as geographer of values.

FROM CITY-STATE TO NATION-STATE

Alasdair MacIntyre, who laments the moral vacuity of modern liberal individualism, has as his ideal a moral life somewhat in the image and likeness of that sketched by Aristotle.[7] He envisages a moral community in which citizens can share a moral consensus.[8] In the background of such laments is the influence of Aristotle's *Politics* and *Nicomachean Ethics* which envisage the ideal political life as that of the citizen in the city-state. The city-state for Aristotle provided the basis for a moral community that could, if properly structured, avoid the relative anomie of the citizen of the contemporary nation-state. For a city-state to provide such a moral matrix, it could not be too large. As Aristotle suggests in the Nicomachean Ethics, "you cannot make a city of ten men, and if there are a hundred thousand, it is a city no longer" (*Nicomachean Ethics* 10.1170 631–32; [3], p. 1091). With a city of proper proportions, governors and citizens know each other well enough to discern each other's characters. In such a context, so Aristotle argues, the proper individuals can be elected to office and lawsuits can be decided correctly (*Politics* vii. 4 1326 310–18). However, when the city becomes too large, one can no longer maintain the moral character that sustains it. Not only will governors (and citizens) not know each other well enough to judge each other's characters, but they will be unable to control immigration which will dilute the character of the state. As Aristotle put it, "When the population is very large, they are manifestly settled at haphazard, which clearly ought not to be. Besides, in an overpopulous state foreigners and metics will readily acquire the rights of citizens, for who will find them out? Clearly then, the best limit of the population of a state is the largest number which suffices for the purposes of life, and can be taken in at a single view" (*Politics* vii.4 1326 618–25; [4], p. 1284). It is in terms of such ideals that the classic Greek philosophical view of the good political life was framed.

It is ironic that Aristotle, who framed a view of politics and ethics based on the model of the city-state, was the tutor of Alexander the Great, who forged the largest Greek nation-state in history. Aristotle was looking to the past as Alexander moved towards a future in which nation-states would embrace communities of men and women with differing moral and religious perspectives. Many of our difficulties in understanding moral life within a nation-state stems from our commitment to an Aristotelian perspective which draws its fundamental images from the model of a city-state. The Aristotelian perspective brings us, however, falsely to hope that the nation-state can embrace and affirm a particular view of the good life. It makes us hope for the impossible, that the nation-state could be *a* community. Indeed, one might suspect that this is in part the root of Alasdair MacIntyre's disquietude with the modern situation. The modern context is one in which the city-state has become but a moment of the larger life of an anonymous, and from the city-state citizen's perspective, amoral nation-state. The Aristotelian political viewpoint has left us ill-equipped for the task of negotiating our moral lives within a nation-state, which if it is not to be a tyranny of one community over the rest must be a secular pluralist state.

Though there are major dissimilarities, one can with benefit note the similarity between MacIntyre's view of the ideal city-state and ideal portrayals of 19th century New England towns in which citizens and their governors could presume mutual familiarity. Such offered a basis for that sound moral judgment which Aristotle thought to be essential to the good political life. In its terms, one can sketch a view of medical practice where generally shared moral viewpoints would allow physicians and patients to meet, understanding each others' characters and share a consensus concerning the goals of medical practice. A form of friendship could direct such encounters of physicians and patients based on shared commitments to goods and purposes in health care. It is in this sense that I have characterized the encounter of physicians and patients from the same moral community as the encounter of friends. They can share together a view of the meaning and purposes of medicine and of the nature of health. In this one can agree with Edmund Pellegrino's reading of Plato in the *Lysis*[9] that physician and patient are friends in pursuit of the good of health (*Lysis* 217). However, again the presumption is that a community of understanding of goods and purposes exists.

The physicians in the nation-state, however, encounter patients who do not share with the physician the same moral community. The physician is like the citizen of a city-state having to deal with metics and foreigners. He or she must encounter individuals who, from the perspective of the physician's own moral commitments, are barbarians. Consider a devout Christian physician from a conservative religious community encountering a patient who is a committed homosexual, a woman seeking an abortion for reasons of convenience, or a family wishing to discontinue treatment for a severely deformed neonate whom the physician holds to be salvageable. The hope to meet as friends, as Edmund Pellegrino characterized the physician-patient relationship, can be fulfilled only if one construes friendship in a very abstract and contentless fashion which approaches what is shared by strangers in the individualistic liberal state taken to task by Alasdair MacIntyre. They meet as strangers in the very straightforward sense that their views of the good life and purposes of health care are very likely to appear strange and exotic to each other. They treat each other peaceably as a condition for the peaceable community, not out of a common commitment to a concrete moral viewpoint.

The very weaknesses of the physician-patient relationship in the nation-state are also its strengths. Though it cannot presume a common view of the goals of health care on the part of physicians and patients, it must presume a commitment to mutual toleration.[10]

The moral language of a secular pluralist state offers a solution to the encounter of individuals belonging to different moral communities, strewn across a nation, if not a world, for it eschews as much content as possible and reflects as far as possible the general conditions for the possibility of a peaceable community, a community based on mutual respect (the remaining source for moral authority). Though there is lack of moral content, there is not in fact anomie. One finds oneself as a person with at least rights to be recognized by bureaucracies in general and by physicians in particular.

The general moral perspective of life and medical practice in the nation-state would, however, lead to a moral vacuousness if one were to suppose that its neutral moral language exhausted the moral world of the citizens of such a state. One must recall that the moral language of secular pluralist societies is framed precisely to allow individuals to speak across communities. It presumes that they continue to live within particular moral communities with their concrete understandings of the goods of life and the purposes of health care. Physicians and patients must, then, live on two moral tiers, one, that of their particular moral community, as well as that of the general secular society available to all individuals in a liberal individualistic nation-state: a state where respect of freedom is a constraint upon political actions, and authority is derived from the consent of those involved.

SUMMARY

The physician-patient relationship must be appreciated within two complementary moral contexts: that of general secular society and that of particular moral communities. In each context the significance of the relationship is determined by the amount of moral content that can be established and the scope of authority that can be claimed with justification. As one would expect, one purchases generality at the price of abstractness, and content at the price of parochiality. The more one is able to reach peaceably across to strangers in the physician-patient relationship, the more abstract the substance of the relationship becomes. The more one is able to reach across to patients as friends in the physician-patient relationship, the more the relationship becomes dependent upon particular views of the goals and purposes of life and health care. The physician-patient relationship is usually a little bit of both and rarely all of one. To look only at one element or dimension is to give a misleading account, at least insofar as one lives in a pluralist society. When one lives and practices medicine fully within the embrace of a moral community, one can not ignore, save as an intellectually important possibility, the ethic of secular pluralist societies. In such societies as ours, the moral viewpoint of particular communities presupposes that of the secular society to give connection with others, while that of the secular society presupposes the moral viewpoints of the particular communities to give the lives of patients and physicians content and purpose. We are in different contexts, both strangers and friends.

Center for Ethics, Medicine, and Public Issues,
Baylor College of Medicine, Houston, Texas

NOTES

[1] As an index of this circumstance, one might consider the major force of works in bioethics framed explicitly within a Judeo-Christian viewpoint, such as [22].

[2] For an excellent overview of some of these issues, see the recent study by Darrel W. Amundsen ([1], [2]).

[3] This presupposition is the fundamental fabric of a minimum notion of ethics as an alternative to force. If one cannot frame ethics in terms of a particular concrete moral viewpoint, one can still understand ethics generally as a means of resolving disputes regarding the probity of alternative avenues of action without recourse to force against the innocent, but rather through appeals to reason, and to the interests of the parties involved.

[4] Here I borrow the notion of respect of freedom as a side constraint from Robert Nozick ([21], pp. 30–34). Unlike Nozick, however, my argument has similarities to a Kantian transcendental argument. I am suggesting that respect of freedom as a side constraint functions as a necessary condition for the minimum notion of ethics and of the peaceable community. Unlike Kant, however, I am presuming a distinction between freedom as a side constraint and freedom as a value. For a further elaboration of my argument in this regard, see ([8], [7], [10]).

[5] A *locus classicus* for the argument that scientific generalizations in medicine are historically, socially, and psychologically conditioned, is the very influential work of Ludwik Fleck, *Entstehung und Entwicklung einer wissenschaftlichen Tatsache* [13]. This book of Fleck has, through its influence on Thomas Kuhn's *Structure of Scientific*

Revolutions [14], had an immense influence upon the literature in the history and philosophy of science. See also, ([24], [16], [23]).

6 If one cannot discover a general content for ethics, one can at least reject those moral maxims that would be incompatible with the minimum notion of ethics as an alternative to force. Only through such a minimum notion can a general framework for blame, praise, and moral authority be sustained. Moreover, such a moral viewpoint requires for its basis and authority, interest only in resolving disputes in ways other than those based directly on force.

7 See, for example, Alasdair MacIntyre's discussion of the similiarities and dissimiliarities of his account with that of Aristotle ([15], pp. 183–185; also, p. 241).

8 Alasdair MacIntyre decries the fact that "modern society is indeed often, at least in surface appearance, nothing but a collection of strangers, each pursuing his or her own interests under minimal constraints" ([15], p. 233). He qualifies this description by rejecting bureaucratic individualism, and places himself within the Aristotelian tradition ([15], pp. 238–245).

9 As such, the bureaucracy embodies as abstract and ahistorical moral perspective as possible so as to span with the least tension the disparate moral communities it embraces. As a consequence, if one were to term the ethos of the secular pluralist state that of secular humanism, one would need to distinguish this sense of secular humanism as far as possible from particular humanist movements. See ([7], [9]).

10 Here one finds the abstractness of the Kantian viewpoint, which was criticized by Hegel (see, for example, *Philosophy of Right*, Section 135), now taken as a virtue. In response to such an Hegelian criticism, the abstract formalism of bureaucratic individualism is advanced in this article as only a moment of the moral life of the nation-state. So appreciated, it is the state acting in its universality. This is suggested, for instance, in Hegel's treatment of civil servants as a universal class (see *Philosophy of Right*, Section 303).

BIBLIOGRAPHY

1. Amundsen, D. W.: 1981, 'Casuistry and Professional Obligations' (Part I), *Transactions and Studies of the College of Physicians of Philadelphia* 3 (March), 22–39.
2. Amundsen, D. W.: 1981, 'Casuistry and Professional Obligations' (Part II), *Transactions and Studies of the College of Physicians of Philadelphia* 3 (June), 93–112.
3. Aristotle, *Nichomachean Ethics*, W. D. Ross (trans.), in R. McKeon (ed.), *The Basic Works of Aristotle*, Random House, New York, 1941, pp. 927–1112.
4. Aristotle, *Politics*, B. Jowett (trans.), in R. McKeon, *The Basic Works of Aristotle*, Random House, New York, 1941, pp. 1113–1316.
5. Burns, C. R.: 1977, 'American Medical Ethics: Some Historical Roots', in S. F. Spicker and H. T. Engelhart, Jr. (eds.), *Philosophical Medical Ethics: its Nature and Significance*, D. Reidel Publishing Co., Dordrecht, Holland, pp. 21–25.
6. Code of Federal Regulations 45 CFR 46.
7. Engelhardt, H. T., Jr.: 1982, 'Bioethics in Pluralist Societies', *Perspectives in Biology and Medicine* 26 (Autumn), pp. 65–78.

8. Engelhardt, H. T., Jr.: 1980, 'Personal Health Care or Preventive Care: Distributing Scarce Medical Resources', *Soundings* **63**, 234–256.
9. Engelhardt, H. T., Jr.: 1982, 'Secular Humanism and Contemporary Bioethics', presented at the University of Illinois Medical Center, Chicago, Illinois, April 21.
10. Engelhardt, H. T., Jr., with M. Malloy: 1982, 'Suicide and Assisting Suicide: A Critique of Legal Sanctions', *Southwestern Law Review* **36** (November), 1003–1037.
11. Engelhardt, H. T., Jr.: 1983, 'Viability and the Use of the Fetus', in W. B. Bondeson, H. T. Engelhardt, Jr., S. F. Spicker, and D. Winship (eds.), *Abortion and the Status of the Fetus*, D. Reidel Publishing Co., Dordrecht, Holland, pp. 183–208.
12. Flathman, R. E.: 1982, 'Power, Authority, and Rights in the Practice of Medicine', in G. J. Agich (ed.), *Responsibility in Health Care*, D. Reidel Publishing Co., Dordrecht, Holland, pp. 105–125.
13. Fleck, L.: 1935, *Entstehung und Entwicklung einer wissenschaftlichen Tatsache*, Benno Schwabe, Basel; 1979, *Genesis and Development of a Scientific Fact*, T. J. Trenn and R. K. Merton (eds.), F. Bradley and T. Trenn (trans.), University of Chicago Press, Chicago, Illinois.
14. Kuhn, T.: 1962, *Structure of Scientific Revolutions*, University of Chicago Press, Chicago, Illinois.
15. MacIntyre, A.: 1981, *After Virtue*, University of Notre Dame Press, Notre Dame, Indiana.
16. McCullough, L. B.: 1981, 'Thought-Styles, Diagnosis, and Concepts of Disease', *The Journal of Medicine and Philosophy* **6** (August), 257–261.
17. Miller, L. J.: 1980, 'Medicine and the Law: Informed Consent' (Part I), *Journal of the American Medical Association* **244**, 2100–2103.
18. Miller, L. J.: 1980, 'Medicine and the Law: Informed Consent' (Part II), *Journal of the American Medical Association* **244**, 2347–2350.
19. Miller, L. J.: 1980, 'Medicine and the Law: Informed Consent' (Part III), *Journal of the American Medical Association* **244**, 2556–2558.
20. Miller, L. J.: 1980, 'Medicine and the Law: Informed Consent' (Part IV), *Journal of the American Medical Association* **244**, 2661–2662.
21. Nozick, R.: 1974, *Anarchy, State, and Utopia*, Basic Books, New York.
22. Ramsey, P.: 1970, *The Patient as Person*, Yale University Press, New Haven, Conn.
23. Sadegh-zadeh, K.: 1981, 'World 5 and Medical Knowledge', *The Journal of Medicine and Philosophy* **6** (August), 263–270.
24. Trenn, T. J.: 1981, 'Ludwik Fleck on the Foundation of Medical Knowledge', *The Journal of Medicine and Philosophy* **6** (August), 237–256.

ALBERT R. JONSEN

THE THERAPEUTIC RELATIONSHIP: IS MORAL CONDUCT A NECESSARY CONDITION?

The question posed in the title of this essay would engage the interest of philosophers. Any serious attempt to answer it, as it is stated, requires skill at the clarification of concepts, like 'therapeutic' and 'relationship', adeptness at logical argument in demonstrating 'necessary conditions', and awareness of the normative and metaethical complexity involved in defining 'moral conduct'. This is worthy work, but I shall, for the most part, refrain from it. Moreover, the way the question is posed leads toward considerations more theoretical than practical. The most philosophically pleasing result would be a clearer understanding of the meaning of the question and of the sorts of arguments which might allow it to be answered in some meaningful way. Again, this is valuable but, to some extent, unsatisfactory. It is the answer, not the ways of answering, which engages the interests of an audience different from the philosophers: namely, those who are related in the therapeutic relationship – physician and patient.

This essay is written for that audience rather than the audience of philosophers. I shall take an approach which skirts some of the genuine conceptual and logical problems raised by the title question (although I hope I do not forget them). For example, I shall not be as attentive to the logical niceties of 'necessary and sufficient condition' as philosophical protocol would dictate. I shall be much bolder in using the terms 'moral', 'morality' and 'moral conduct' than current critical discourse in moral philosophy might tolerate. Finally, 'Therapeutic Relationship' will be described in a manner far more empirical than one usually encounters in philosophy of medicine. The goal of the essay will be to propose an answer to a question actually being asked, i.e. 'Of what importance to good medicine is good morality?'

One prominent practitioner, Eric Cassell, states his conviction about this question:

> I believe that medicine is inherently a moral profession. . . . The practice of medicine – caring for the sick – takes what are presumed to be facts about the body and disease and on the basis of that technical knowledge does something for a *person*. In that sense it can be seen in the same light as any moral behavior – moral because it has to do with the good and welfare of others ([2], p. 87).

Earl E. Shelp (ed.), The Clinical Encounter, 267–287.

Louis Lasagna, another physician, takes a quite different position. Responding to the claim that medicine is a 'moral enterprise', he remarks,

> Is medicine a 'moral enterprise'? I doubt it, any more than plumbing or auto repair is. Physicians rarely cure or save lives, and spend most of their time trying to provide some comfort, relieve symptoms and perhaps prolong life. The moral issues seldom come up in situations that constitute physician's practice. The dramatic life and death issues . . . are difficult, but relatively uncommon occurrences for the typical physician ([13], p. 44).

However, he goes on to state that there are many occasions on which a physician may act immorally: avoiding the dying patient and drunks or preferring personal convenience to patient's needs. How then, is Lasagna's position different from Cassell's? Only, it appears, in that Lasagna (misled perhaps by the grandiose phrase, 'moral enterprise') presumes the morality of medicine to be concentrated in the wrenching dilemmas and critical decisions that ethicists delight in analyzing. Yet he and Cassell both see morality and immorality in the daily activities of medical practice. For both proponent and critic morality is important to medicine.

The importance of morality in medicine is not, of course, a modern idea. It runs as a theme through the history of Western medicine. Admittedly, it is a minor theme in a literature which fills enormous tomes with technical information and observation; yet the assertion of the importance of morality is never lost. Justification of this importance is neither extensive nor common, but several different arguments can be discerned. First, for many centuries, the medicine of the Western world lived within a strong religious tradition, in which the physician was considered an instrumental cause of healing utilized by the Primary Cause, the Divine Healer. In this Judeo-Christian perspective, the physician must be worthy to be so used. Personal holiness was a prerequisite to being an effective instrument.

Physicians' prayers, composed in the latter era of this great cultural tradition, express this sense of unity between personal holiness and medical efficacy. A sixteenth century prayer reads, "God, eternal Father . . . sanctify, guide and direct my undertakings through the Holy Spirit so that I may accumulate beneficial knowledge and apply it successfully to the healing art. Thou alone art Author and Source . . . of successful cures and all healing is from thee." The famous Prayer of Moses Maimonides, actually composed by the Jewish physician Markus Herz in the 18th century, asks: "Support me, Almighty God, in these great labors (as a physician) that they may benefit mankind, for without Thy help not even the least thing will succeed" ([7], pp. 36, 29).

A second sort of justification proposes that the character of the physician is, in some sense, part of the treatment. The Hippocratic Corpus provides one of the earliest statements about the importance of morality to medical practice. It foreshadows the modern conception of physician as 'placebo'.

If there be an opportunity of serving one who is a stranger in financial straits, give full assistance . . . for where there is love of man there is also love of the art. For some patients, though conscious that their condition is perilous, recover their health simply through contentment with the goodness of the physician (*Precepts* VI).

The famous phrase, "where there is love of man (philanthropia), there is love of the art", although it has a noble ring, probably does not refer to the virtue of altruism, but merely to friendly amiability. The author is advising physicians that their benevolent and cordial behavior can be of help to their patients. This advice too, repeated throught medical history, anticipates the modern conception of the physician as 'placebo' [6].

A third justification is a quite common sense one. It is offered by a famous Portugese physician of the 17th century, Roderigo à Castro:

There are many excellent physicians who, lacking (the principal virtues of the physician, generosity and prudence) render themselves unwelcome to their patients and ultimately useless [12].

The virtues of the physician are, in this view, characteristics which attract the patient and make the physician 'welcome'. Physicians who fail in these virtues will be useless to their patients, simply because their patients will be disinclined, and even refuse, to use their services.

The tradition of Western medicine, then, reveals several different justifications for the importance of morality. The first, exemplified in the quotations from the physicians' prayers, requires personal holiness in order to be a worthy instrument of God, Primary Cause of Health. The second, suggested in the Hippocratic text, claims that moral behavior, described as 'love of man', will act as a therapeutic influence. The third, in à Castro's words, proposes morality as a useful means of maintaining contact between patient and physician.

All of these traditional justifications are interesting. The first, preaching the importance of personal holiness, is not exactly a modern theme, but its human appeal should not be discounted. However, it is seldom claimed today that the efficacy of medicine requires each practitioner to be an Albert Schweitzer or a Tom Dooley. The second claim, the therapeutic influence of the physician's morality, does have a modern echo. W. R. Houston wrote an essay in 1938 entitled 'The Doctor Himself as Therapeutic Agent' [10], in which he describes certain characteristics, such as concern and compassion, which might be considered 'moral characteristics'. These, he says, strengthen the relationship between patient and physician in ways which have therapeutic effect. Subsequent authors have frequently noted that the physician is the placebo, and that physician behavior which encourages trust and hope is vital to therapy. Some attempts have been made to demonstrate this thesis empirically, but without much success: yet it seems intuitively true and verified by experience. The third justification is also a current one: physicians who lack the characteristics considered by this society as morally fitting ones will lose their patients, sometimes to the patient's detriment. This thesis seems plausible, but only if it is assumed that the patients believe the physician's 'immorality' will affect them. There does appear some willingness to tolerate behavior which patients do not consider any of their business. Many patients continue to frequent a doctor whom they know to be gamblers, philanderers, tax cheats (behavior which is commonly but not universally branded immoral). Still, the essential point of à Castro's claim is true: if a patient departs from a physician's care because the patient is offended by the physician's immorality, that physician's technical proficiency is rendered useless for that patient.

The traditional justifications for the importance of morality are interesting. However, the contemporary question of great interest today is the definition and measurement of 'competence' in medical practice. The question about morality corresponds to this interest. Does moral behavior have any definite relationship to medical competence? Is morality an integral component of competence? Can a physician be judged to be competent and immoral at one and the same time? If there is any definite relationship, can it be described in ways which allow its presence or absence to be discerned or evaluated? At a recent meeting of all Specialty Boards devoted to the subject of evaluation of competence, one report fascinated the audience: a study of a group of pediatric resident physicians revealed a significant positive correlation between level of moral judgment (as measured by the Kohlberg scale) and the degree of clinical skill. This report, while more stimulating than conclusive, focuses the question about morality in medicine: can it be shown that good moral conduct makes for better medical performance [19]?

This question might be answered by some sort of empirical research. But, obviously, before such research could be designed, the relevant terms must be defined. What is one to look for when told to observe 'moral conduct?' What activities count as medical performances and how are they measured?

The formidable task of defining and designating the content of moral conduct intrigues philosophers but baffles concerned physicians. I wish to suggest that 'moral conduct' can be integrated into the components of clinical competence so that its importance can be seen as necessary to competent clinical practice. Moral conduct, as so described, is not an adjunct to good medicine, but intrinsic to it: without such moral conduct medicine cannot be good.

Dr. Edmund Pellegrino, a physician-philosopher, has written, "... ethics in medicine is not a relative piety adopted by its practitioners, but a necessary and productive basis for action" ([17], p. 49). He proposes a reconstruction of medical morality in which the act of profession of the physician as healer and the competence of the physician are inextricably linked to moral conduct. I accept Dr. Pellegrino's suggestion on the relationship between competence and moral conduct. While my route will differ somewhat from his, it will end up at same place: affirming that the central act of medicine is the 'right and good healing action for this patient in these circumstances'. It is 'right and good', not merely in the technical sense but in the ethical sense. The technical sense takes these words only to mean that the physician's acts conform to certain scientific theories, logical criteria and empirical data which lead, for example, to 'prescribing the correct antibiotic for the particular organism'. The ethical sense adds to this the claim that the acts are right and good because they flow from certain virtues, conform to certain moral principles and are, in their essence, an act of moral discretion.

THE THERAPEUTIC RELATIONSHIP

Much has been learned about the relationship between physicians and patients from sociological models (cf. [16], [9]). While no one model appears sufficient in itself to explain the complex nature of this relationship, each provides significant insights into its dynamics and into the place of that relationship within the wider social structure. This essay will describe the relationship in its own way, drawing to some extent on the insights of sociological models, but with a different perspective and a different interest. The different perspective is that of the moral philosopher rather than the social theorist and the interest is in answering the question posed in the essay's title rather than in describing the social dynamics. In this light, the therapeutic relationship will be viewed in four perspectives. These perspectives are (1) the 'moral archeology' of the profession of medicine, (2) the expectations of the practitioner and of the patient, (3) the structure of the practices central to medicine, (4) the impact of the activities of the relationship on the surrounding social environment. Each of these dimensions will be described and the moral elements which are involved in it stated. The relevance of moral conduct outside the therapeutic relationship for the relationship itself will be explored. The result, it is hoped, will be an affirmative answer to the question posed in the title of the essay. It will not, however, be an answer justified by a single argument, but by many complex and diverse considerations.

THE MORAL ARCHEOLOGY OF THE PROFESSION OF MEDICINE

Medicine is a social institution with a long history in Western culture. Those who participate in it today, either as its practitioners or as actual or potential patients, cannot avoid perceiving it in the light of certain beliefs which flow out of that history. Very few of its participants are medical historians, but all carry within their understanding of medicine some fragments of that history understood as a moral archeology of medicine. They understand, to some extent, what it is to be 'doctor' and 'patient' and what 'medicine' does, because of those fragments.

That archeology, like most real archeological remains, is complex, but two lineaments stand out clearly: Western medicine can be understood in terms of two of its historical realizations, Greek medicine and early medieval medicine which carry quite different moral messages. As we understand Greek medicine today, it consisted of a collection of skills, based on certain beliefs about nature, which were taught by practitioners to apprentices. Practitioners offered their skills to patients in return for compensation. Greek medicine was, to the best of our knowledge, a mercantile enterprise which, as we saw in the Hippocratic text quoted above, made exceptions for those who couldn't pay the full price. But, we see little evidence that the practitioner felt any strong compulsion to serve the poor and the needy: he was a craftsman with a product to sell and he had to make a living by it. There was no moral opprobrium in so doing [6].

Early medieval medicine was inspired by the Christian message of love for the most abandoned. Quite commonly, monks were its practitioners. Despite its technical and scientific impoverishment, it seems to have taken seriously the admonitions derived from the example of Jesus who healed the blind and the lame and the example of the Good Samaritan, described by Jesus, who, on encountering a man beaten by robbers "bound up his wounds . . . brought him to an inn and took care of him", even paying for his room and board. Service to those in need of care, regardless of their ability to pay, became an ideal constantly preached throughout the medieval era [20].

These two moral lineaments persist in our cultural beliefs about doctors. The physician has attained a skill at great cost and has the right to employ that skill to earn a living. The physician's skill is often most needed by those who cannot pay and he has a duty to provide it out of charity. Both principles can be maintained but, clearly, they will sometimes come into conflict. Even more important, their persistence creates an inner tension in the institution of medicine. Should its activities be ordered primarily for charity or primarily for profit; should need or ability to pay be the foundation on which the structures of medicine are erected? Is the profession to be viewed as a collection of entrepreneurs who are particularly benevolent or as a group of persons whose skills oblige them to serve the neediest and the poorest?

Whether one is a physician or a person who occasionally needs a physician, these beliefs shape the relationship. Physicians believe they have a right to earn a living and that they do have some special obligation to those who need their skills. Patients acknowledge that the doctor should be paid, yet feel they deserve care when they cannot pay. A morality of legitimate self-interest and of altruism coexists in the moral archeology of the profession. Although taken for granted in a general way, this dual morality causes tension in the structure of medical care, in the behavior of physicians and in the expectations of the public. The fact that they can often be reconciled in particular cases by 'making arrangements' does not alleviate the tension.

Social institutions are not 'things' out there in the world, but certain practices and arrangements of people about which those people have beliefs. Among those beliefs will be some which are of moral import. Persons approach each other in the institution with these moral understandings about it and each other. In medicine, persons encounter a physician in light of these understandings which are, of course, colored in different ways for different people and in different settings. Judgments are made by individuals and by the public about 'good' and 'bad' doctors with these understandings in the background.

For the purpose of this essay, the two lineaments of altruism and self-interest have been emphasized. This is done, first, to show they both have a traditional legitimacy and, second, to note that they can, while comfortably coexisting for the most part, come into conflict. It is a first step in answering the question posed in the title to state that the therapeutic relationship exists only with this dual moral archeology beneath it. Conduct which reflects the lineaments of altruism and self-interest is the substance and substratum of the therapeutic relationship.

THE EXPECTATIONS OF THE PRACTITIONER AND THE PATIENT

Expectation (if a new metaphor can be introduced) is the fuel of activity in a social institution. The institutions move and are energized by the expectation of the participants. Wherever they are located in the fact-value debate, they are undeniably important [18]. In medicine, practitioners and patients encounter each other with certain expectations. Some of these are derived from the moral archeology of which we spoke above. Many other expectations are brought from the current perception which persons have about the function and capability of medicine and from the personal beliefs and experiences of the participants. Many of these expectations are idiosyncratic: a person may approach a doctor with unreasonable beliefs about the physician's powers to abolish all evils; a physician might deal with patients with a personal sense of infallibility. However, a wide range of expectations can be assumed to be common: persons come to doctors because they feel in themselves a certain malaise which is culturally believed to represent a malady about which doctors can do something. They submit themselves to the doctor's doings in expectation that the cause of the malaise can be discovered and removed. Similarly, doctors accept patients because they expect the patient to submit to them a malaise and to respond to their attempts to remove its cause. Another set of expectations refers to the bearing which the doctor should exhibit to the patient and vice versa. The patient expects the doctor to be attentive to the complaint and, perhaps, even to be sympathetic; the physician expects the patient to be open and cooperative.

What is the moral import of these many expectations with which physicians and patients encounter each other? Certainly, if these expectations become the matter of contract or promise, they are incorporated into a familiar moral practice. This does happen, to some extent, in the so-called 'fiduciary contract' which the law presumes as the legal framework of the therapeutic relationship. The patient expects the physician to act 'in his best interests'; conversely (although not a part of the legal doctrine), the physician expects the patient to consider 'his best advice' and 'to do his best' to be helped. But the expectations of both parties have a moral import in themselves. The expectations which both parties bring to the encounter are, at the beginning, somewhat vague and ill defined. As the encounter continues, the expectations are shaped and new expectations emerge.

Expectations inevitably shape people's behavior. In a relationship, expectations must be expressed in ways which allow each participant to respond. Failure to express expectations will introduce distortions into the relationship and eventually dissolve it. Distortions take place when new forms of behavior are prompted by changed, but unexpressed, expectations, which the other partner can no longer understand nor respond to appropriately. For example, the patient may be disappointed in his expectation of cure and become gradually non-compliant with medical advice; the physician, no longer expecting to be able to provide notable relief, may become disinterested in the patient and intolerant of his visits. It is crucial to the maintenance of a therapeutic relationship that expectations be expressed as they change and grow in the course of events.

It seems easy to say, 'expectations must be expressed', but there must exist a determination in both participants which tolerates and encourages that expression. In recent years, moral philosophers have made much of the notion of respect. Introduced by Immanuel Kant to describe the moral principle whereby each person should be treated as an end and not as a means merely, it has been elaborated to encompass the recognition by each person that he and all others are autonomous beings, the center of personal choice and value. The notion is philosophically complex, yet it strikes a chord in our culture. Indeed, many moral philosophers suggest that respect is the foundation of morality, since acknowledgement of the other as a being of independent value appears to be a necessary presupposition of all moral discourse [4].

Respect, then, can be considered as the principle which allows a relationship to exist in an evolving fashion. As events reinforce or modify expectations, each person becomes aware and is enabled to respond to these expectations in terms of their own views and values about life. It is the attitude which creates the atmosphere in which people come to know what they can expect from each other. Lack of respect ignores the expectations of the other, cuts off their expression, distorts them into the 'means' that will serve one's own ends. Since a relationship is a continuing encounter with changing expectations, respect is the necessary condition for its existence.

Physicians do meet patients in various ways. The sustaining power of respect functions differently in each sort of encounter. The single encounter can barely be called a 'relationship', a word which denotes at least some continuity and intensity. Yet, even here, respect functions to elicit the nature of the complaint, to calm fears, to communicate instructions. Even the brief encounter, if it is to be therapeutic, requires respect. As encounters extend into periodic, episodic and chronic meetings, they take on the characteristics of a relationship and require a stronger and deeper sense of respect for sustenance of the relationship.

Respect is a fundamental moral principle. It is much more profound than the colloquial use of the word suggests (a use which hints at mere tolerance at worst and bemused indulgence at best, e.g., one should have respect for one's elders or one's teachers). At the same time, out of respect as a fundamental moral principle flow a variety of qualities and attributes which should be exhibited in the physician-patient relationship, e.g., sensitivity, concern, attentiveness, absence of bias and prejudice. These attitudes, qualities and attributes are psychological characteristics: persons possess them to a greater or lesser degree and manifest them in different ways, depending on genetic heritage, upbringing, cultural experience and their highly individuated personalities. Often, they are not easily recognized and are sometimes disguised. These qualities, which are often mentioned when the moral attributes of the physician are discussed, certainly have moral import, but they are derivative from the fundamental moral principle of respect. The former are variable in degree and manifestation; the latter is constant and resolute, the same in all persons who adopt 'the moral point of view', namely, that the other person is an independent, valuing and valuable being [1]. It is this fundamental moral principle which makes possible the relationship and which is the condition for it to become 'therapeutic'.

THE STRUCTURE OF THE PRACTICES OF MEDICINE

We have, until now, spoken of the relationship primarily as an ongoing and sustained encounter of two persons; we have not stressed the therapeutic feature which marks this relationship as one between a physician and a patient. The word 'therapeutic' can be used in a broad sense, designating that two persons encounter with the expectation that one is a healer and that the other needs or wishes to be healed. The term may also be used in a restricted sense (as it often is in psychiatry) to propose that the relationship itself is a healing one. I shall use it in a sense intermediate to these two: the therapeutic relationship is an ongoing and sustained encounter in which the practices constitutive of medicine are performed by one person and enacted upon and by the other. The practices constitutive of medicine are multiple, but two central ones can be designated as properly constitutive of medicine as we know it today in the Western world: diagnostic and therapeutic activities performed by the doctor and enacted upon or by the patient.

The essence of diagnosis is the successive formation and testing of hypotheses in the light of emerging data and theories of disease. The process is aimed toward a specific end point, the discovery of the nature of the disease process which is troubling the patient and, in some instances, its cause. Therapeutic activities flow from a clinical judgment that the diagnosis reveals a disease which is or is not amenable to certain treatments: drugs, surgery, counseling, instructing and, in some cases, even waiting. Therapeutic activities are enacted 'upon and by' the patient because in some the physician alone is active and the patient passive, such as in a surgical operation, in others the physician and patient are both active, as when the one prescribes and the other takes a medication, and in some others, the patient alone is active on the basis of the doctor's advice, as in giving up smoking or taking up exercise.

These two constitutive practices of medicine are, of course, found in multiple manifestations. They are sometimes not properly diagnostic nor therapeutic in the proper sense of the words: in some instances, no diagnosis can be made but the patient continues to be 'followed' by the physician; in certain instances, therapeutic activities do not 'cure' but do alleviate symptoms, provide support for impaired function and, in some cases, bring about effects which appear unrelated to the therapeutic intervention.

These constitutive practices are, in modern medicine, quite technical. However, they are more than technical behaviors; they are interactions between persons, and, as such, have a moral dimension. It is a serious mistake to think of these activities in terms of a model like drawing blood, growing a culture, isolating an organism and, on the basis of these steps, making the diagnosis: infection by *Salmonella typhi*. This is a marvelous achievement, but it is only a part of the diagnostic practice. Presentation of a complaint by the patient, physical appearance and signs, conversation about recent events, etc., all surround the precise technical steps. In essence, diagnosis is the revelation of a person to another person. The facts and information so revealed must be understood and interpreted. The exchange and interpretation of information is a human practice. It is governed by human intentions, motives and emotions. Since this information exchange has a specific goal, namely, the discovery of the nature and cause of malaise, it must be kept on the track by a dominant intention in both patient and physician. This dominant intention, directing the information exchange and excluding deviation and deviousness, is a moral quality called honesty.

Human practices are done well or poorly not merely because human powers are capable of error, but because they can be effected by the motivations of the performer. Thus, all human practices, including the technical ones of diagnosis and therapy, have their virtues, that is, a direction to a goal imposed by the willing choice of the performer. As information is elicited from the patient, it must be structured and ordered by the mind of the practitioner and viewed in relation to a scientific theory of disease: nothing can be omitted because it is inelegant or incoherent, nothing can be added as more satisfying or more logical. The data must be gathered as accurately as possible, marshalled as fully as possible and a logic applied to it which comes not merely from the physician's preference but from the contemporary science of diagnostics. All of this requires utmost honesty, for all of it has to be directed to the goal of discerning what is the patient's disease. Dishonesty implies that the information could be utilized with another end in view: to lead the patient into unneeded surgery, to terrorize or dominate the patient. Honesty is the moral dimension of the practice of diagnosis. Thoroughness and carefulness are the minor virtues of the diagnostic procedures which flow from the major virtue of honesty.

It is obvious that honesty must also be a virtue of the patient. The history must be accurate, the complaint reflect a true perception, the desire for help genuine. Devious or untrustful patients distort, even if they may not totally destroy the diagnostic process. Honesty is a virtue mutually required of both participants in the therapeutic relationship.

The second of the constitutive practices, therapy, also has a moral dimension, but it is somewhat less easy to discern than honesty in the diagnostic process. Therapeutic activities flow from a clinical judgment which mediates between them and diagnosis and which is defined as the "ability of the physician to choose wisely and selectively among alternative possibilities and includes the determination of the appropriate balance of risk and benefit of alternative courses of action or inaction" ([3], p. 16). A wise and selective choice would appear to be one which followed upon an evaluation of the probability and magnitude of the risks and benefits entailed by any course of action or inaction. A wise and selective choice is a courageous one. In short, the moral dimension in therapeutic activities is the virtue of courage.

Courage sounds pretentious: old eulogies of doctors praised the 'courageous' Dr. Rush who risked life to fight the yellow fever. It also sounds prejudicial: is it the doctor or the patient who must run the risks? The physician will not die nor suffer and will, in the end, be paid regardless. Despite its pretentious and prejudicial tone, courage would seem to be the right word for the moral dimension of therapy. Aristotle offered the classic definition of courage: it is the "measure of feelings of confidence and of fear, that is, the ability to maintain judgment against the excess of rashness or the deficiency of timidity in matters of danger and of great moment" (*Ethica Nichomachea* III 1116a10). Admittedly, some clinical judgments are safe enough and the results matter little. But for many medical judgments, much is at stake, sometimes death, more often, debility and discomfort. Physicians who make such clinical judgments cannot be timid nor can they be rash: they must always habitually walk the narrow path marked caution-confidence. This does not mean that some physicians will incline somewhat to one side or the other of that path. It does mean that the physician who is habitually timid will forego many benefits for patients; the one who is habitually rash will endanger them to no good purpose. Courage is the moral dimension of therapy.

Of course, as we mentioned above, the physician advises about the risks and performs the activities that entail them, but does not, in fact, encounter them. The patient does. The patient, too, needs courage in making judgments about what he or she wishes to undertake. It may be a courage that differs from the physician's. Nevertheless, timidity and rashness incur the same sorts of penalties. The courage of the physician is not a virtue that faces the danger of personal death and disability, as it is for the patient. It is rather a virtue that faces the danger of loss of pride in oneself as a healer, loss of reputation and, above all, failure to respond to the expectation of one who has entrusted himself to care. The physician must, in these ways, allow himself to be vulnerable 'in matters of danger and of great moment'. He must step out from behind the defenses of hesitation and shun the coward's retreat into indecision and inaction. There are baser dangers as well: such as loss of income, but it can be presumed that the courageous physician is also primarily moved by the principle of respect that measures success and failure in terms of altruism rather than self-interest.

The structure of the two practices constitutive of medicine require a moral dimension: in one, the virtue of honesty, in the other, the virtue of courage [14]. Neither diagnosis nor therapy will reach their goals without these virtues in those who enact them or in their recipients. These virtues are the necessary conditions for their efficacy, although the conditions of technical skill and informed and correct judgment are also required. Needless to say, neither practice may be efficacious, since both are liable to the 'medical uncertainty' of which Renée Fox has eloquently written [8]. Still, the moral virtues and the technical skills operating together should lead to the 'right and good healing action', an action that, all other things being equal, will in fact heal. Thus, two specifically moral virtues, honesty and courage, are necessary to the therapeutic relationship, viewed in terms of its constitutive practices.

IMPACT ON THE SOCIAL ENVIRONMENT OF THE RELATIONSHIP

The therapeutic relationship can be seen as an isolated social phenomenon, taking place between several persons over a stated period of time. However, every therapeutic relationship has an impact on the wider world in which it takes place. The totality of therapeutic relationships has significant impacts. These impacts are multiple: they have to do with the health status of the nation and of particular populations, employment of persons in the health professions and the health care industry, financial investments and costs, educational enterprises. The macro-world of health care is made up of the micro-world of the therapeutic relationship. In recent years, this macro-world has been described in considerable detail and efforts made to trace the causes and effects that flow between it and the micro-world. The incidence of surgical procedures, for example, have been traced to the number of surgeons which has been traced to the educational and the economic system. It appears that the constitutive practices of diagnosis and therapy are influenced to a considerable extent by the macro-world and vice versa.

As a result of the containment of every practice within this microworld, there is a problem of justice within any particular physician's practice. Every therapeutic relationship is but an instance in a number of actual and potential relationships. Each physician has a finite number of patients at any one time and must spend a finite amount of time and energy with each. Similarly, each person (except in emergency situations) determines the time, energy and money they wish to spend on medical care. Decisions must be made about allocations of time, energy and resources, both by the practitioner and the patients, and these will involve the moral issue of 'fairness'. On what grounds is one patient to be given a greater share than some other or one patient entitled to take more than another.

This sort of problem is familiar to the medical tradition of Western culture. The moral archeology of the tradition allows a place for a morality of self-interest beside a morality of benevolence. The balance between the dual moralities has been struck by insisting on a principle of justice: the physician should not look to any distinguishing characteristic of patients other than the patient's need for medical care. In one striking expression of this principle, inscribed on the Temple of Aesculapius, the 'god-likeness' of a physician lies in following a principle of justice as impartiality:

> Physicians must be like gods: Bringing benefit to all alike, the rich and the poor, men and women, friend and enemy.

In essence, the distribution of benefits based upon merit, desert or entitlement is excluded in favor of a distribution on the basis of medical need [15]. In modern times, physicians, even without knowing the tradition, deal with the problem of justice in their own practices by relying on an allocation of service based upon medical need. A contemporary physician would be unlikely to devote time to a bothersome patient with no serious medical need to the detriment of a seriously ill patient. Thus, the virtue of justice, which distributes benefits in accord with an appropriate principle, is a characteristic of the therapeutic relationship.

There is also a major problem of justice facing the therapeutic relationship. Each physician controls a significant amount of the resources (it is said that every graduate from medical school adds $500,000 to the costs of medical care each year). The manner in which these resources are employed in any particular relationship has an almost invisible, but cumulatively great, impact upon medical care as a whole. The ways in which individual physicians and groups of physicians organize the provision of care, as well as the styles of their practice, can improve or restrict access to future patients as well as enhance or reduce service to present patients.

These questions introduce into medicine a moral problem seldom discussed in earlier times: the problem of social justice. Do certain forms of medical practice and health care create social situations in which some people can obtain all desired care and others are deprived even of needed care? If so, does this constitute an injustice? Can and should this injustice be remedied? These are significant matters for social policy. Do they have any relevance for the therapeutic relationship? In one sense, it is obvious that they do. Certain forms of therapeutic relationship may be altered or eliminated by social policies aimed at a more just distribution of health care. For example, the forms of that relationship which involve a personal physician might be abolished or those relationships designed for certain sorts of care, such as psychotherapy, might be curtailed. But does the problem of justice have any place in the therapeutic relationship as we know it?

It has been noted that physicians bear some responsibility for the 'medical commons' [11]. At the same time, those who note this responsibility hesitate to suggest that a physician formulate clinical judgments about particular patient's needs in light of social needs. Certainly, some 'cost containment' can be effected by improvements in diagnostic skills and use of less expensive, though equivalent drugs. The intrusion of social considerations into clinical decisions has been viewed skeptically: leave such considerations to policy makers, it is said, not to clinicians. Still, the separation of powers cannot be so clean. Can clinical decisions about the availability of renal dialysis abstract from criteria such as social usefulness and productivity of the potential patient? Can clinical decisions about coronary artery bypass surgery or about neonatal intensive care be made in isolation from information about the economic effects of these procedures and their social utility?

In the past, physicians justified high fees for the rich by providing free care to the poor: in so doing, they acknowledged the problem of justice. Today, choices of specialization or styles of practice should be made, not only in light of personal preferences, but in light of the effect on a just distribution of care. Thus, it can be said justice is a moral imperative in the therapeutic relationship, both in its immediate demands of fairness within a practice, and also in its broader, more vague mandates to practice a form of medicine which promotes rather than impedes a more just distribution of the benefits of care. Long ago, a physician writing about medical ethics made a statement that, although it had a different meaning in his philosophical world, can be recalled in a more modern sense. The Renaissance physician, Giovanni Codronchi, wrote, "a physician may have all other virtues, yet in lacking justice, lose all other virtues" ([12], v. 3, p. 954). This may be taken to mean that even the intrinsic virtues of the therapeutic relationship, respect, honesty and courage, will have little effect if the therapeutic relationship is not available to those who need it when they need it. Unless the relationship is justly distributed, for many persons it will not exist at all.

PRIVATE BEHAVIOR

This essay proposes that the therapeutic relationship is built out of the moral archeology of the profession, the structure of its practices, the expectations of its participants and the socio-economic climate. Each of these elements has been shown to have a moral dimension. Up to now, we have concentrated on the moral characteristics of the relationship itself. Another question can be raised: does the moral conduct of persons apart from the relationship have any relevance to the relationship itself?

The question now becomes, 'must the good doctor (one who has both technical skills and the virtue proper to the therapeutic relationship) also be a good *person* (exhibit morality in all aspects of personal life)?' This is obviously a very difficult question. First, there is very little agreement about what 'morality' involves in general or whether certain acts are moral or immoral. Secondly, human life is not an unchanging continuum but a series of events woven together in complex patterns. The 'moral person' is not a rigid statue, but a growing, changing, responding being, caught in tragedy and in triumph, seizing opportunities and losing gambles. In the midst of this, a person who lives by certain principles which might commonly be considered 'moral' may perform an immoral act or fall into some immoral habit. What does it mean to speak of moral conduct or moral persons? Finally, until now, this essay has suggested that the moral conduct proper to the therapeutic relationship is incumbent upon both participants: physicians and patients alike must be respectful of each other, honest, courageous and fair. Can we now reasonably suggest that both must exhibit moral conduct in their daily lives in order to maintain a therapeutic relationship? In what sense must the patient be a moral person in order to be a good patient?

There are major obstacles in the way of discussing this question. They can be reduced somewhat by considering only those forms of behavior which would commonly be considered immoral, by concentrating only on the physician side of the relationship and by pointing out a peculiarity in the philosophical analysis of moral conduct. No definitive answer to the question will be provided, given all these obstacles and limitations, but some ruminations that appear plausible, to me at least, will be offered.

Recent newspaper stories provide examples of conduct by physicians that would, I believe, be widely branded as immoral (and not only because they are also illegal). In one case, a physician convicted of one rape and accused of two others was employed by a hospital. In a second case, a physician was convicted of Medicaid fraud and, as soon as he served his (reduced) prison term, returned to medical practice and to the practice of defrauding insurance carriers. A third case relates that a physician defaulted on his medical school loan while keeping up payments on a luxury automobile.

In the case of the convicted rapist, a doctor, although aware of this incident, wrote a recommending letter: He later explained, "I gave him a medical reference because . . . I believe he's a good doctor . . . (the rape) is absolutely not relevant" (*Boston Globe*, Sept. 22, 1981). A hospital official said, "although this is not a pleasant situation, it does not impact on (his) ability to deliver medical care to patients" (*Boston Globe*, Jan. 23, 1981).

These responses seem to me intuitively implausible and, indeed, outrageous. Why should such statements appear plausible and unobjectionable to anyone, as apparently they did to the recommending doctor and to the administrator? There are, I think, two reasons. First, 'medical care' and 'good doctor' are taken in a narrow sense that views the practice and the practitioner in exclusively technical terms. Medical care is taken to be a set of standardized, routinized and validated procedures; the practitioner is seen as one who has acquired the knowledge and skills to employ the procedures effectively. Technical skills must, of course, be artfully applied, but why should they be morally applied? We have tried to show that even technical performance cannot be built into a relationship apart from certain moral principles and virtues, but there is no reason why the performance itself cannot be well done without any reference at all to morality. However, there is also no good reason to refer to the disparate performance of technical acts as 'medical care' nor to the performer as a 'good doctor'. These terms clearly denote much more than skillfully accomplished disparate performances. They refer to the entire context in which the performance is done and to the relationship. Thus, it is possible that the most skilled performer of prostatectomies be a bad doctor and that the most accurate diagnosis be part of very bad medical care whenever the moral principles and virtues proper to the therapeutic relationship are absent. We often say, jokingly, "if it's my prostate coming out, I'll take the bad guy with the best record." But, we know, in our more serious moments, that we don't want him for *our doctor* any moment longer than the operation.

Still, this narrow interpretation of 'medical care' and 'good doctor' and its refutation does not answer the question. The obstacle to the answer is not so much a narrow understanding of medical activities, but rather a narrow understanding of moral conduct. Many persons untutored in moral philosophy would readily accept the proposition that a rapist, a cheat and a freeloader would not be a good doctor. Moral philosophers, however, might have more trouble with the proposition. This stems from the recent preoccupation of moral philosophers with a very narrow view of moral conduct. For several decades, moral conduct has been broken into individual discrete 'actions' and the question asked, "what characteristic of this action makes it permissible to designate it as 'morally good' or 'morally bad'?"

A few voices of protest are now being raised against this narrow view [14]. The protestors complain that the view fails to take into account what almost all persons do take into account in moral judgments, namely, the character of the one performing the actions. If there were no modern moral philosophers to confound the issue, we would speak readily of a 'good or a bad person'. We would probably not be very clear about the logic or the empirical grounds for the statement, but we would be clear that it referred to the habitual and characteristic ways in which an individual presented himself to the world and to companions. Even individual actions would often be evaluated in the light of character: we might, for example, be more inclined to excuse or to seek some justifying reason for an untruth told by one whom we know to be generally truthful than we would be for a known liar.

Character refers not only to experience of past habitual behavior of persons but to anticipation of future performances. The honest person is more likely to be trusted in a tempting position, the courageous one in a dangerous job. We may be proven wrong and our trust disappointed, for actions take place in individual circumstances which cannot by fully predicted. At the same time, we make important judgments about the future of activities and enterprises on the basis of judgments about the character of participants in them. Character refers, then, to the self as known in past action and projected into the future. It refers to the whole person as one whose motives and intentions in all phases of life can be reliably, if not infallibly, assessed and predicted.

If this view of character, rather than the narrow view of action, were acknowledged as the fundamental factor in moral conduct, it would be quite plausible to suggest that a physician's conduct outside the therapeutic relationship would be relevant to the relationship itself. The relevance would be the likelihood that a character exhibiting certain strengths and weaknesses in general life would exhibit them in the relationship itself.

If this seems plausible, we may take one further step. In the opening pages of this essay, we spoke about the moral archeology of the profession, noting that it contained a dual morality of self-interest and altruism. The duality is legitimate, but precarious. Might it not also be plausible to suggest that the person whose character is flawed by disrespect, dishonesty and injustice (as well as other broadly censured characteristics) would be inclined to favor only one side of the dual morality? The delicate balance would be tipped to the self-interest end; its pans loaded up with considerations which, given the values embraced by the measurer, would easily outbalance the considerations in favor of altruism. Even if self-interest is disguised as altruism (as any clever scoundrel quickly learns to do), the choices would consistently go against one of the partners in the relationship. That partner would, of course, be the weaker, more vulnerable one, the patient. Since a considerable imbalance of power in favor of the physician exists even in the most appropriate therapeutic relationship, the patient is at double jeopardy in the hands of an immoral physician.

These concluding thoughts are, as I said, mere ruminations. I have used words such as 'likelihood' and 'plausibility'. I have pointed to negative features, such as the tendency to view medical practice and practitioners in technical terms and the absorption of moral philosophers with action rather than character. I have not employed positive arguments which could clearly demonstrate the proposition that the bad person cannot be a good doctor. I have not addressed the very complex philosophical problems about whether the virtues are one or many (a problem which lurks behind many statements in this essay) nor whether the virtues pertain to roles alone or to character as well [14, 5]. Finally, I have no empirical evidence, collected in a randomized trial or in any other way, on which to base my ruminations. Yet, is not the demand for irrefutable logical argument and strong empirical evidence a symptom of our own flawed view of the nature of moral discussion? Aristotle, in the First Book of his great treatise on ethics, *The Nichomachean Ethics*, proposed, "our discussion (about ethics) will be adequate if it has as much clearness as the subject matter admits of, for precision is not to be sought alike in all discussions" (*Ethica Nichomachea* I, i, 1094b 14). This discussion, it can be admitted, may need more clarity, but it should not be expected to provide an ineluctable demonstration. Yet, in the absence of such a demonstration, one will continue to hear, 'we have no proof', 'we cannot be certain' and, in the absence of an elusive certainty, some bad persons will be admitted to the profession of medicine, allowed to continue within it and even be protected by it. Patients will be physically harmed and morally offended. The profession will be maligned. The hope is that, even without irrefutable proofs and overwhelming evidence, many good persons will accept the moral task of becoming good doctors, and will, by their respect, honesty, courage and justice, manifest the necessity of moral conduct for the therapeutic relationship.

School of Medicine, University of California,
San Francisco, California

BIBLIOGRAPHY

1. Baier, K.: 1965, *The Moral Point of View*, Random House, New York.
2. Cassell, E.: 1976, *The Healer's Art*, Lippencott, Philadelphia.
3. *Clinical Competence in Internal Medicine*, 1979, American Board of Internal Medicine, Philadelphia.
4. Downie, R. S. and Telfer, E.: 1970, *Respect for Persons*, Schocken Books, New York.
5. Downie, R. S.: 1971, *Roles and Values*, Methuen, London.
6. Edelstein, L.: 1967, 'The Professional Ethics of the Greek Physician', *Ancient Medicine*, The Johns Hopkins Press, Baltimore.
7. Etziony, M. B.: 1973, *The Physician's Creed*, Charles Thomas, Springfield.
8. Fox, R.: 1980, 'Vision and Reality: The Evolution of Medical Uncertainty', *Milbank*

Memorial Quarterly/Health and Society **58**, 1–28.

9. Friedson, E.: 1970, *The Profession of Medicine: A Study in the Sociology of Applied Knowledge*, Dodd, Mead, New York.
10. Huston, W. R.: 1938, 'The Doctor Himself as Therapeutic Agent', *Annals of Internal Medicine* **11**, 1416–1420.
11. Hyatt, H.: 1974, 'Protecting the Medical Commons: Who is Responsible', *New England Journal of Medicine* **293**, 235–238.
12. Jonsen, A. R.: 1977, 'Medical Ethics, Europe in the 17th Century', *Encyclopedia of Bioethics*, Vol. 3, The Free Press, New York, pp. 954–956.
13. Lasagna, L.: 1977, 'Discussion of Do No Harm', in S. Spicker and H. T. Engelhardt, Jr. (eds.), *Philosophical Medical Ethics: Its Nature and Significance*, Reidel, Boston.
14. MacIntyre, A.: 1981, *After Virtue*, Notre Dame University Press, Notre Dame.
15. Outka, G.: 1974, 'Social Justice and the Right to Health Care', *Journal of Religious Ethics* **2**, 11–32.
16. Parsons, T.: 1951, *The Social System*, Free Press, Glencoe.
17. Pellegrino, E. D.: 1979, 'Toward a Reconstruction of Medical Morality: The Primacy of Profession and the Fact of Illness', *Journal of Medicine and Philosophy* **4**, 32–56.
18. Searle, J.: 1964, 'How to Derive Is from Ought', *Philosophical Review* **73**, 43–58.
19. Sheehan, T. J., *et al.*: 1980, 'Moral Judgment as a Prediction of Clinical Performance', *Evaluation and the Health Professions*, pp. 393–404.
20. Sigerist, H.: 1941, *Medicine and Human Welfare*, Yale University Press, New Haven.

DAVID H. SMITH

A THEOLOGICAL CONTEXT FOR THE RELATIONSHIP BETWEEN PATIENT AND PHYSICIAN

INTRODUCTION

I have been asked to discuss the influence of religious ideas on the doctor-patient relationship – a massive and complex assignment. It is massive because the gods worshipped by humankind are as diverse as Apollo and Dionysus, Baal and Ahura Mazda; and it is complex because the actual or ideal influence of religion is notoriously hard to explicate. Since Max Weber, it is notorious that religious beliefs may have social implications different from the rules or virtues they explicitly defend, and there is no agreement among theologians about the possible or proper relationships between traditionally based and rational ideals, principles and forms of human community.

Rather than enter into these difficulties, my strategy is a simple one. I shall focus on the largest complex of religious traditions in the West – the Christian traditions – and I shall attempt to sketch their implications for the relationship between physician and patient with a very broad brush. I shall not attempt to assess whether the relations I sketch *should* obtain, nor shall I cite historical data to establish the actual behavior of physicians and patients in the Christian West. Rather, I want to suggest the kind of conceptual world in which Christianity locates the therapeutic relationship, for I think it is the constitution of that world that is decisive for the doctor-patient relationship as we know it. While I shall focus on Christianity, I believe that most of what I shall say is true for Judaism as well.

MONOTHEISM

Christianity and Judaism share a tradition in which a central claim is that there is only one God. Our culture is so permeated by this claim that we seldom consider the intellectual possibility of polytheism. This is unfortunate, as most persons in the present and past have been polytheists. In any case, for our immediate purposes the important thing to note about monotheism is its negative side: If there is only one God, then the gods of household and fertility, tree and war are not true gods. As Karl Barth once put it, "Olympus and Valhalla decrease in population" [7] at the proclamation of

Earl E. Shelp (ed.), The Clinical Encounter, 289–301.

the sole sovereignty of the one God. And there is a continuous tradition in Western theism that is much surer about what God is *not* than about what God is.

Whatever the problems thus created for the religious conscience, and they are considerable, an important aspect of this emphasis is to desacralize both the political and the natural order [2]. This shift was epistemologically and ontologically significant. Polytheism implies a chaotic universe, one in which various aspects of the world work according to independent principles. Indeed, in a certain sense there is no *uni*verse, only a set of interrelated worlds of crops, household, health and the king. Monotheism unified the world under the power of the one God, and in doing so it suggested the coherence of the world in its relationship to this one power.

Ontologically united, the monotheistic universe is also knowable. The taboo loci of polytheistic ritual are moved out of the historic and natural world. There have been complications in this history. Superstition has always remained a part of the monotheistic traditions, and they have not always been willing to follow an argument where it leads. They have equivocated in their support for the development of modern science, and they have not unequivocally supported the study of anatomy or the development of scientific medicine. But monotheism makes scientific study acceptable in a way polytheism could never do, for the natural correlates of polytheism are the taboos of the many sanctuaries.

Monotheistic religion in the West has had a more particular characteristic, however; it is associated with the idea that God is the 'creator' or 'maker' of the world. Among historians of religion it is widely acknowledged that this is a derivative idea – the religion of ancient Israel began with the Exodus and Lord of History rather than with the Maker of Heaven and Earth. But derivative or not, this assertion that God is the cause of the world is characteristic of Western monotheism; it reinforces the general tendency of monotheism per se; and it is the backbone of the dominant tradition of medical ethics in our culture.

In the earliest myths, the natural world created by God was pronounced to be 'good'. The natural and the good were synonymous. Health was understood to be both a moral and a physical good because of the intrinsic value of anything created by God. Furthermore, the claim that the natural world is a good creation of God provides a theological warrant for the idea that the study of nature can yield morally binding rules or principles. This claim about justifying moral principles with reference to the natural structure of things is, within Christianity and outside it, highly controversial. Its most persuasive formulation remains the thirteenth-century work of Aquinas. Of course a naturalistic ethic without theistic roots is possible, but the symbolism of creation that lies behind this version of Western naturalism has been most influential in the relationship between physician and patient.

These symbols have a powerful legacy, for they tend to suggest that an expert on the study of the proper physical function of a thing will also be an expert on the right moral use of the thing. We tend to expect botanists to be able to tell us which plants should be grown, and we assume that automobile mechanics are especially qualified to tell the difference between good and bad cars. Analogously we think students of the human body naturally learn something about what human beings *ought* to do. This is because we assume health to be a moral value, and that assumption in our culture has a religious root. The whole idea that the objective of medicine is the pursuit of health, understood to be the natural good of the body, derives from this monotheistic claim [5]. The objective of the therapeutic relationship in this context is not ultimately determined by the sovereign will of a patient, but by the prior act of God, who created a natural world that is good.

This affirmation of the goodness of the physical world has always caused religious and intellectual problems, nowhere more acutely than in medicine. If health is the ideal, disease, deformity, sickness and death are the reality. How is it possible that a good God could cause these evils to happen, or allow them to happen?

Paul Ricoeur has brilliantly sorted out layers of response to this question in *The Symbolism of Evil* [13]. For now, it is important to note that the dominant tradition has explained the existence of evil in the world through human moral evil or sin. Whose fault is the tumor on my leg? The Western tendency has been to avoid putting the blame on God. It is my fault, or my father's fault, or it exists for the sake of the character it will make me develop. In general this stance suggests an activistic and moralistic response to disease – even if the activity encouraged is initially moral rather than scientific. An interesting effect is to de-escalate the importance of the therapeutic relationship, even as regards health care, for these theories imply that the *raison d'être* of disease is not physical at all, but that health is achieved or earned through moral righteousness.

Indeed, this moralism is another characteristic component of the traditions of ancient Israel. The whole movement of prophetic protest had as its theme the idea that God's fundamental concern was with justice and righteousness. All natural relationships – domestic, economic and political – were subjected to *moral* judgment. Thus, healing or salvation was not simply a ritualistic or priestly matter but fundamentally involved the self and its social relations. The prophet was at least as much a healer as the doctor. A priestly monopoly over healing seems to have been reasserted in the early middle ages as Laín Entralgo points out [6], but the differentiation of function has been maintained for the most part, with liberating consequences for medical research and practice. In the moral monotheism of the West, the doctor was not a priest and neither was he the ultimate healer.

CITIZEN MENTALITY

The dominant metaphor in the Western religious imagination is life within the kingdom of God. The omnipotent God is understood to be a king, with human-kind as the citizens of His kingdom. To be a Christian or a Jew is to be such a citizen. There are variations on the theme to be sure – the city of the Davidic monarchy is not the kingdom of God of Jesus's proclamation, or Augustine's *civitas dei*, or the realm of brotherhood of the social gospel. And other metaphors have supplemented and displaced this one [9]. But the notion of membership in a community ruled by God comes up again and again in the traditions of the Abrahamic faiths.

A kingdom in which God is sovereign is one in which human sovereignty is limited. The rightness of an act is determined by its conformity to the will of the reigning sovereign. To have that will in written form as law and moral rules is a great advantage. Sometimes in some of the Western traditions it has been assumed that people could intuit God's will directly. In general the epistemological claims have been more modest, and rules for Christian or Jewish life have been *inferred* from a limited body of religiously endorsed traditions: scripture, Talmud or magisterium. As we have noted, there is an ongoing tradition of supplementing these revealed sources of moral knowledge with some kind of natural law.

However they are derived, these rules, regulating all aspects of human life, provide a characteristically nonteleological cast to a portion of Western religious ethics. Their existence implies that there are some things that can be done that should not be done – perhaps even some things good in themselves which are not to be done. Thus there are moral limits on the kinds of therapy we should perform, even for the sake of health.

Within the theocracy any human relationship is a relationship under law. This is true of family, economic and political relationships and it became true of the therapeutic relationship as it evolved in the West. Logically this need not have been the case. We can imagine a therapeutic relationship unhampered by outside constraints, one in which anything the patient's health might require – or anything the patient might request – was to be done. Some may find such a relationship ideal, but it is not the kind of therapeutic relationship we have known in the West, where medicine is less important than morality. For both doctor and patient it is better to suffer evil than to do evil. Several aspects of this morally structured therapeutic relationship call for comment.

A central and relevant rule from the decalogue to the present has been a prohibition on killing the innocent. Of course it has been clear for centuries that killing might be in the interest of the physician, the family or even the patient. But medical killing, with only a few significant exceptions, has been prohibited by religiously colored ethics. Some aspects of the prohibition on killing are particularly interesting. The Christian Church, for example, began with a nearly exceptionless formulation of this rule. But Augustine argued that there was one kind of situation in which killing might be justified: when it was necessary to protect the life of an innocent third party. Self defense was not the paradigm of justifiable killing; defense of another was. If I can save a victim by killing the striking murderer, then kill him I may – indeed, *must* – do.

This justification for killing is very different from a common one in our modern sensibility. It says that I may kill killers: soldiers or criminals. In a relativistic age we are uncertain of our judgments about whether such persons should be killed. Capital punishment and war are morally problematical for many of us. At the same time, we are confident that there are some lives that are not worth living. We speak of mercy killing, therapeutic abortion and rational suicide. It is a one hundred eighty degree reversal of sensibility. The traditional religious ethic, formative on the therapeutic relationships we know, had no room for killing someone in virtue of his diminished capacity. Killing was justified only in terms of something one had done, was doing, or – perhaps – might do. The key notions were threat or desert, not *value* of life, and the religious ethos of traditional medicine ruled out judgments based on the quality of life.

It may be instructive to spend a little time on one specific issue on which this consciousness of citizenship in the kingdom of God runs counter to the modern sensibility. This issue is the question of suicide.

The Bible contains neither an explicit word for suicide nor an explicit prohibition of the act. In the Hebrew Bible Saul and his armour bearer fall on their swords (I Samuel 31:3–5); Ahitophel hangs himself after the failure of his political intrigue (II Samuel 17:23); and Zimri burns himself to death after the failure of his attempt to take the throne of Israel (I Kings 16:18f.). Samson prays to die with the Philistines he kills as he topples their temple (Judges 16:23–31). In the New Testament, Judas is presented as a suicide (Matthew 27:4–5; cf. Acts 1:18). Yet in none of these cases is there any suggestion that the persons are bad, or fail to live up to obligations, *because* they had committed suicide. The moral quality of the acts of Saul and his armour bearer is at worst ambiguous; Samson's act seems meritorious.

On the other hand, one can see in these documents bases from which a prohibition on suicide could be inferred by later Jewish and Christian writers. If suicide is the appropriate form of death for Ahitophel, Zimri and Judas, it is so because of the seriousness of their moral wrongdoing. Both the sixth commandment (Exodus 20:13) and the penalty for the shedding of human blood (Genesis 9:5–6) suggest an opposition to suicide. So does the notion of God's sovereignty over human life found throughout the Bible but especially in such passages as Genesis 1–3, Romans 14:7–12, I Corinthians 6:19.

This theme of the incongruence between God's sovereignty and suicide can be found outside the Western religious tradition. In Plato's *Phaedo* (61–62), Socrates concedes that there are times when a person would be better off dead. Why not act oneself to bring about this state of affairs? The prohibition of such action is based on the fact that persons are 'possessions' of the gods. Just as I would be wronged by the theft of one of my possessions (such as my cattle), so the gods are wronged when one of their possessions is taken. Personal authority over one's own life is limited by the prior sovereignty of the gods; suicide represents a usurpation of power.[1]

Whatever the correct reading of Plato may be, the idea that human beings have limited sovereignty over their own lives finds a natural home in the teaching of the Western religions. Developing Jewish traditions condemned suicide, taking as the key text not the prohibition on murder in the decalogue but that on shedding human blood in Genesis 9:5f. Sacrifice of one's own life was only allowed as a last resort to avoid committing one of the cardinal sins of adultery, murder or idolatry. Death was better than betrayal of the name of God (and thus many medieval European Jews died rather than be forced into apostasy) but it is God, not persons, who is the Lord of Life and Death. A Jew may not take his life to avoid profaning the sabbath; the burial ritual for suicides is greatly abbreviated. Similarly in Islam, despite the *Koran's* omission of an explicit statement on the subject, Sura 3:139 ("It is not given to any soul to die, save by the leave of God, at an appointed time . . .") has been taken as the grounds for a prohibition on suicide. Reinforced by *hadith* telling of the Prophet's refusal to bury someone who had committed suicide, this prohibition was included in the *Sunna* of the community.

Martyrdom, respected by Judaism, Christianity and Islam, seems inconsistent with this general theological prohibition. The debate over the issue in Christian moral theology is instructive. Against certain philosophical and fanatical Christian tendencies Augustine argued that the Biblical prohibition on killing when "rightly interpreted" includes no exception "in favor of him on whom the command is laid" ([10], p. xx). The prohibition of killing is not exceptionless for it does not protect plants and animals who are "dissociated from us by their want of reason", it does not exclude just action by public agents. The possibility of a "special commission" from God is admitted as in the case of Samson who received "secret instructions" from the Holy Spirit ([10], p. xxi). Yet these exceptions do not amount to a justification of suicide. Only the state is to execute the guilty ([10], p. xvii), and the concession about special divine instructions is meant to cover only a few cases ([10], p. xxvi).

Thomas Aquinas claimed that suicide is not only an offense against self and society but also a violation of God's sovereignty. People belong to God, their creator, and suicide is somewhat analogous to theft. A thief steals my property and in suicide I steal God's. "Because life is God's gift to man . . . whoever takes his own life sins against God, even as he who kills another's slave, sins against that slave's master. . . ." (*Summa Theologica* II/II q. 64 A. 5).

Several twentieth century theologians have based their opposition to suicide on a revised version of this theological argument. They argue that suicide is wrong because it is an inappropriate response to God's *gift* of life. Suicide is an act of ingratitude, a failure to recognize that God is the Owner of human life ([3], p. 402). In the context of Western monotheism, they insist, it is nonsensical to speak of a 'right to die', although the rightness or appropriateness of death can be acknowledged. The sovereignty of patient and physician is displaced and limited by the sovereignty of God and the general rules of His kingdom.

At least one other characteristic of moral life in the monotheistic theocracy has been very important for the relationship between patients and physicians: the pluralism of basic duties or rules. While Jews and Christians often have defended the notion of love (or some equivalent) as an ordering or supreme principle, the plurality of duties and conflicts to which that leads has had to be faced as a practical matter. Even if doing good and telling the truth are both demanded by love, there are times in which it appears one must choose between these two firm requirements of the religious conscience. An ongoing challenge for religious moralists is choosing a strategy for handling this problem.

For example, a person dying slowly and with unrelievable pain may request to be 'put out of his misery'. The theocratic moralist cannot resolve this issue simply in terms of the patient's choice, as I have suggested. No more sufficient is appeal to the obviously fundamental requirement of doing good to others, for the tradition knows an apparently independent prohibition on killing. Thus Christian and Jewish moralists have faced the dilemma of how to do good without killing. If there were an individual and absolute right to die, that would solve the moral problem. If beneficence were the sole rule in the theocracy, the issue would disappear. But neither of these has been the confession of church or synagogue. The result has been to complicate enormously reasoning about medical moral matters.

Several interesting tactics have been taken to resolve this dilemma. Perhaps the most important has been an attempt precisely to specify the range of personal responsibility. Persons are said to be responsible for what they *intend*, for example, in ways they are not responsible for more remote consequences of their acts. Thus heavy prescription of analgesics is essentially therapeutic. Even if death is hastened it is not thought to be murderous in the way poisoning with arsenic would be. By narrowing the unit to which the rules apply, one is able to hold to the force and pluralism of the rules. Precise definition of action becomes important because of the pluralism of duties, the complexity of our moral landscape.

Western religious ideas, in other words, have structured the therapeutic relationship, limiting both its goals and the means medicine can use to attain those goals. Insofar as those ideas lose credibility we can expect to see some substantial changes in the medical ethos.

SUFFERING AND CHRISTOLOGY

Some of the tendencies described are reinforced by a complex of attitudes toward suffering characteristic of Christianity. In Buddhism suffering is transcended; in Christianity the emphasis is different. The major ideas come out clearly if we examine the major Christian theodicies. John Hick [4] has helpfully delineated two types. The dominant Augustinian tradition was picked up by the medieval church and Aquinas. It controls in the thought of both Luther and Calvin and has been resurrected in modern dress by Karl Barth. The general lines of this theodicy are familiar and clear enough.

According to it, evil is the product of a primordial fall by heretofore perfect creatures (either angels or humans). The evils of the real world are fundamentally punitive for that 'original' crime. How could an omnipotent and good God have allowed this to happen? This is the narrow question of theodicy and it is, as Henry David Aiken once remarked [1], a function of the 'monotheistic syndrome', i.e., it is easily avoidable if one is willing to alter one's beliefs so as to accept a finite God, make God's goodness equivocal, or deny the reality of evil.

The common Western response to this problem is the so-called free will defense in which the characteristic Western preoccupation with freedom is ascribed to the deity. God, so the argument goes, wanted to create beings in His own image. They must, therefore, be free. They would not have been free had He created them without the possibility of sinning, or if, predictably saints, they had had only the *abstract* possibility of choosing an evil alternative. A range of ancillary problems inevitably comes up: What is the origin of the evil impulse in the original being(s)? Did God know and in some sense predestine the Fall? If so, does He predestine redemption as well? Why should God will a creation involving a range of seemingly dysfunctional beings? Interesting as these issues are, I have neither time, ability nor interest to pursue them here.

What is central to our purpose is to note that in terms of the Augustinian tradition (as delineated by Hick), suffering is always in some sense of the word *deserved*. It may not be deserved by the sufferer per se, but if not, then it must be deserved by someone: Adam, all humankind, relatives. Thus a conception of vicarious or substitutionary suffering is central to this view. Related to this understanding of suffering is a conception of God as a judge and, obviously, of the relationship between persons and God as being fundamentally juridical. God is assumed to be the all-just, all-powerful monarch Who treats His human subjects according to their deserts.

Hick himself constructs what he calls an Irenaean tradition that germinated in the thought of the second century Gaulish theologian and has been more systematically developed by 'liberal' theologians beginning with Schleiermacher in the nineteenth century. In Hick's reconstruction of this view the theological distinction between the 'image' and the 'likeness' of God is crucial. Persons are made in God's image, but they must perfect themselves as moral agents in order to live in God's likeness. The Fall is not a key doctrine, in either a literal or metaphorical sense. The germ of truth in the Genesis account is its testimony to the present imperfection of human nature. Evil is necessary as an obstacle to be overcome. The world is a vale of 'soul-making' in which moral personality can develop. For this development to occur, persons must be free: able to choose between good and evil, able to make mistaken judgments about that which is good. This means that they must live at a certain 'epistemic distance' from God, related to Him only by faith. They need space in which to grow into His likeness [4].

Therefore, on this view, suffering serves a fundamentally pedagogical function. As weights, obstacle courses and calisthenics are necessary for the development of the physical self, so pain, regret and disappointment enable the formation of a vigorous, virtuous character. The world is a kind of gymnasium or classroom, and God is the ultimate nondirective teacher, creating an environment in which genuinely self-directed learning and growth can occur. God's ultimate responsibility for evil is admitted 'up front', but this does not discredit Him. There is a sense, indeed a time, in which this evil will be seen to have been worth the suffering. (With Kant, Hick postulates immortality.) Thus Hick is able, in his own way, to affirm the famous *'O Felix Culpa'* (O Blessed Fall), for without moral struggle there could be no moral virtuosity, indeed no character at all.

The connection is not altogether logical, but I think that we can see an affinity between the two theodicies just discussed and our own responses to suffering. The Augustinian tradition suggests that it is appropriate for suffering to exist as a punishment. The natural ethical inference is that suffering that is *deserved* is appropriate; undeserved suffering should not exist and should be fought. The medical result would be to suggest that it is more important to treat infectious diseases, the results of natural disasters and genuine accidents, than forms of suffering that people in some sense bring on themselves. Desert might be related to habitual behavior (like smoking or drinking), carelessness, environmental practices, social habits, etc.

In contrast, the Irenaean tradition's claim that suffering's role in the world is pedagogical, plausibly leads to the idea that our major concern should be with *pointless* suffering. Pain that causes me to move away from destruction, disappointment from which I learn to redirect my energies – those forms of suffering can be said to have a point. Although they may be ameliorated, they do not require us to bring out the heavy artillery. The big guns should be brought to bear on the suffering that serves no good purpose: pain, unhappiness or guilt that enervates and simply destroys. We should grow ourselves as we alleviate the pointless sufferings of others, rather than trying to create the morally unchallenging and therefore unnourishing world that would exist if persons never suffered at all. Were we to adopt this paradigm we would find ourselves constantly making judgments about the 'value' of a specific form of suffering. Will it be good for my friend to endure that? Can he learn from the experience?

There is no doubt that ideas like these have been taught by the churches, or that they have had an effect on the practice of medicine in our culture. Patients are expected, or expect themselves to be, strong enough to endure pain. On learning of another's misfortune we often look for a congruence with previous misdeeds, or for a sense in which he asked for it, or for some good that may come out of the situation. In moralizing the situation, we betray our roots.

At the same time there is another tendency in Christianity, more clearly manifested in art and worship than in doctrine. The crucifix, baptism, the eucharist all suggest the Christian's identification with Jesus – in particular with the suffering Jesus. The symbol of Christ on the cross should be the characteristically Christian datum in debate about suffering, and theory construction is temporally and epistemically secondary. Thus prior to theological speculations about atonement, about the Fall, or about the value of suffering there is a simple insight: Whatever and whoever He may be, at least Christians can be sure that God suffers with us; we do not suffer alone. Christian art and literature through the ages have focused on the suffering of Christ because it was comforting. Why? Because in his own suffering the Christian is identified with God. "When you suffer, *your sufferings are God's sufferings*, not his external work, not his external penalty, not the fruit of his neglect, but identically his own personal woe. In you God himself suffers, precisely as you do . . ." ([14], p. 843).

Sometimes ideas like these have led to an idealization of suffering and indifference to pain and discomfort. When combined with belief in personal immortality they can yield a supernaturalism that is complacent about human misery. But that need not be the result, for the example of the suffering Christ is an example of care for the needy. If the Christological symbolism has sometimes drugged the sensitivities in medicine, at least as often it has inspired selfless service.

BONDING

In general it seems fair to say that the monotheistic, theocratic and Christological themes in Christianity combine to create a world in which the physician knows only one power, has a variety of duties to it, and lives a self-sacrificial life for his patient as God live(s) for him. The fidelity of God to his people is the paradigm for the devotion of the physician to his/her patient. "A professional eats to heal, drives to heal, reads to heal, comforts to heal, rebukes to heal, and rests to heal" [8]. Covenant-fidelity is the core of both Christian ethics *and* the therapeutic relationship according to Paul Ramsey ([11, 12]).

The upshot is to sanction a relationship of total concern for the patient at great cost to the healer and with the serious potential for monomania. I leave it to others to judge the extent to which this is or has been the dominant sociological reality. Christian and Jewish traditions about humility and the love of God, about the order to charity and consolation can provide helpful correctives. Opposition to idolatry runs very deep.

On balance it seems hard to say whether the qualifications introduced into our medical world by religious symbols and traditions are good or bad. The surest thing is that the qualifications are there, affecting our moral consciousness in pervasive ways. Perhaps it is possible to sustain some of the good aspects of this legacy purged of the bad; perhaps it can be done without benefit of tradition, worship and religious community. If we are to avoid discarding baby with bath, we would be well advised fully to explore the possibilities for critical and traditional reconstruction.[2]

Indiana University,
Bloomington, Indiana

NOTES

[1] It is difficult to determine the exact force this argument was meant to possess. Not only does the uncertain status of religious arguments in Plato's moral theory arise, but the precise location of this argument in the *Phaedo* is relevant. It is not presented as an exceptionless argument, since suicide is said to be legitimate when there is a sign from the gods. Socrates clearly thinks that his own actions of self-destruction are not in violation of whatever obligations to the gods he may have. More generally, in the *Phaedo* Plato is concerned to argue that death is not intrinsically evil. Some virtuous persons (philosophers) constantly try to separate things of the soul from things of the body. Their lives are lived in pursuit of death. Bodily life is not absolutely good; virtue may require a distancing from it. Plato's ideas of value and virtue are not congruent with a general prohibition on suicide.

[2] Portions of this essay have appeared in the *Encyclopedia for Bioethics* (s.v. 'Suicide'), Warren T. Reich, Editor-in-Chief (© 1978 by Georgetown University, Washington, D.C.); and in Carol Levine (ed.), *Essays on Death, Suffering and Wellbeing* (forthcoming). I am grateful to The Free Press, a division of Macmillan Publishing Co., Inc., and the Kennedy Institute of Ethics, Georgetown University, and to the Hastings Center for permission to reorganize and reuse this material.

BIBLIOGRAPHY

1. Aiken, H. D.: 1962, *Reason and Conduct: New Bearings in Moral Philosophy*, Alfred A. Knopf, New York.
2. Frankfort, H.: 1949, *Before Philosophy*, Penguin Books, Baltimore.
3. Gollwitzer, H. (ed.): 1961, Karl Barth, *Church Dogmatics: A Selection*, Harper and Brothers, New York.
4. Hick, J.: 1978, *Evil and the God of Love*, rev. ed., Harper and Row, San Francisco.
5. Kass, L. R.: 1975, 'Regarding the End of Medicine and the Pursuit of Health', *The Public Interest* **40**, 11–42.
6. Laín Entralgo, P.: 1969, *Doctor and Patient*, McGraw-Hill Book Company, New York.
7. Mackey, A. T. *et al.* (trans.): 1961, Karl Barth, *Church Dogmatics* **III/4**, T. & T. Clark, Edinburgh.
8. May, W. F.: 1975, 'Code and Covenant or Philanthropy and Contract', *Hastings Center Report* **5** (December), 29–38.
9. Niebuhr, H. R.: 1963, *The Responsible Self*, Harper and Row, New York.
10. Oates, W. J. (ed.): 1948, Augustine, *The City of God* in *Basic Writings of St. Augustine*, Vol. 2, Random House, New York.
11. Ramsey, P.: 1970, *The Patient as Person*, Yale University Press, New Haven.
12. Ramsey, P.: 1978, *Ethics at the Edges of Life*, Yale University Press, New Haven.
13. Ricoeur, Paul: 1967, *The Symbolism of Evil*, Beacon Press, Boston.
14. Royce, J.: 1898, 'The Problem of Job' in John J. McDermott (ed.), *The Basic Writings of Josiah Royce*, Vol. 2, University of Chicago Press (1969), Chicago.

NOTES ON CONTRIBUTORS

George J. Agich, Ph.D., is Associate Professor, Departments of Medical Humanities, Philosophy, and Psychiatry, and Director, Ethics and Philosophy of Medicine Program, School of Medicine. Southern Illinois University, Springfield, Illinois.

Darrell W. Amundsen, Ph.D., is Professor of Classics, Western Washington University, Bellingham, Washington.

Baruch A. Brody, Ph.D., is Professor, Departments of Medicine and Community Medicine, and Director, Center for Ethics, Medicine, and Public Issues, Baylor College of Medicine, and Professor of Philosophy, Rice University, Houston, Texas.

K. Danner Clouser, Ph.D., is Professor of Humanities (Philosophy), College of Medicine, the Milton S. Hershey Medical Center, Pennsylvania State University, Hershey, Pennsylvania.

John Duffy, Ph.D., is Priscilla Alden Burke Professor of History, University of Maryland, College Park, Maryland.

H. Tristram Engelhardt, Jr., M. D., Ph.D., is Professor, Departments of Medicine and Community Medicine, and Member of the Center for Ethics, Medicine, and Public Issues, Baylor College of Medicine, Houston, Texas.

Gary B. Ferngren, Ph.D., is Associate Professor of History, Department of History, Oregon State University, Corvallis, Oregon.

Isaac Frank, Ph.D., is Senior Research Scholar, Kennedy Institute of Ethics and Adjunct Professor, School of Medicine, Georgetown University, Washington, D. C.

Albert R. Jonsen, Ph.D., is Professor of Ethics in Medicine, and Chairman of the Bioethics Group, School of Medicine, University of California, San Francisco, California.

John Ladd, Ph.D., is Professor of Philosophy, Brown University, Providence, Rhode Island.

Joseph Margolis, Ph.D., is Professor of Philosophy, Temple University, Philadelphia, Pennsylvania.

Laurence B. McCullough, Ph.D., is Associate Director, Division of Health and Humanities, Department of Community and Family Medicine, School of Medicine, and Senior Research Scholar, Kennedy Institute of Ethics,

Earl E. Shelp (ed.), The Clinical Encounter, 303–304.

Georgetown University, Washington, D. C.

Edmund D. Pellegrino, M. D., is the John Carroll Professor of Medicine and Medical Humanities, School of Medicine, Georgetown University, Washington, D. C., and Professor of Philosophy, Catholic University of America, Washington, D. C.

Earl E. Shelp, Ph.D., is Assistant Professor of Medical Ethics, Department of Community Medicine, and Member of the Center for Ethics, Medicine, and Public Issues, Baylor College of Medicine, and Associate Professor of Theology and Ethics, Institute of Religion, Houston, Texas.

David H. Smith, Ph.D., is Professor and Chairman of Religious Studies, and Director of the Poynter Center, Indiana University, Bloomington, Indiana.

Robert M. Veatch, Ph.D., is Professor of Medical Ethics, Kennedy Institute of Ethics, Georgetown University, Washington, D. C.

Patricia D. White, J. D., is Assistant Professor of Law, Georgetown University Law Center, Washington, D. C.

INDEX